T0295684

False Prophets of Economics Imperialism

Also by Matthew Watson and published by Agenda

The Market

False Prophets of Economics Imperialism

THE LIMITS OF MATHEMATICAL MARKET MODELS

Matthew Watson

To Katie

First published in 2024 by Agenda Publishing

Agenda Publishing Limited
PO Box 185
Newcastle upon Tyne
NE20 2DH
www.agendapub.com

ISBN 978-1-78821-766-8

British Library Cataloguing-in-Publication Data
A catalogue record for this book is available from the British Library

Typeset by JS Typesetting Ltd, Porthcawl, Mid Glamorgan
Printed and bound in the UK by TJ Books

Contents

Preface

The publication of this book represents the culmination of a very lengthy process. The underpinning research began as part of my ESRC Professorial Fellowship project, Rethinking the Market (grant number ES/K010697/1, 2013–19). I remain extremely grateful to the ESRC for providing the funding to allow me to think my way into a brand new project and for giving me the time to see this book through to its natural conclusion. This has enabled me to educate myself in a number of fields that previously I knew next to nothing about. In particular, I have spent many fascinating hours discovering new leads in the history of mathematics, the history of metamathematics, the history of science and the history of hypothetical scientific modelling, before then trying to work out how they should influence my understanding of economics imperialism. As a consequence, the scope of my argument has extended far beyond where I would have been able to locate it at the start of the process. The result, I hope, is a study that self-consciously operates in the interstitial spaces between different disciplines, different traditions of thought and different bodies of work. There would have been little point in labouring hard for a decade only to have ended up where the existing literature was already situated. The synthesis of various intellectual registers is a deliberate attempt to chart new ground.

It is not entirely straightforward to enter the debate about economics imperialism while avoiding the overtly partisan tone of the existing literature. Not all of it falls into the same trap, but typically analysis from first principles is forced to give way to judgemental assertions from those who never doubt that their position has been the right one all along. Respectful invitations to discuss issues that might always have the ability to polarize therefore often get drowned out by raised voices. My readers will ultimately decide whether I have managed to stay away from the tendencies I regret in the writing of others. Even having selected the title of *False Prophets of Economics Imperialism* might be seen by some as a provocation too far. However, at least in my own mind I have concentrated on the multiple ways in which arguments might be made in the social sciences and what role the twin processes of mathematical proof-making and mathematical modelling might have in explanation. I am especially interested in the epistemic justification for mathematical market models, because this is what economics imperialists carry with them when entering other social scientific domains. My focus is on how the process of argumentation might proceed in this way, how the ensuing arguments navigate the essential ontological distinction between the world within the model and the world beyond, and whether they live up to the often grand claims made in their name. I am content to leave more divisive position-taking to others.

However, I am aware that there is always going to be more at stake in writing this book than solely seeking the limits of the application of market models in the explanation of non-market social phenomena. Merely to ask this question is to stir the pot, to challenge readers to think about where they stand on the issue. There will be no lack of opinions in what follows, and some readers will assume they have found more smuggled into the text than those I am actively willing to endorse. This comes with the territory of writing about as evocative and as controversial a topic as economics imperialism. Wanting to focus wherever I can on analytical matters does not mean I have repressed the urge to consider what counts as a step too far. Clearly I have my own views on where the limits of these mathematical incursions might best be placed, and they will not be to everyone's liking. In the past I have also often managed to divide my audience, where economists have accused me of a thinly veiled exercise in economics bashing and non-economists have bemoaned my failure to launch a more devastating attack. Neither of these things was my intention, then or now, but I accept that different readers will have different responses to what I have written and different thoughts again about what I should have written. This, though, is how debates start, and therefore I am comfortable with other people reading the pages that follow in the way that makes most sense to them.

If I do have a specific message in mind, it is for those people in political science, international relations, law, sociology, cultural studies and history who are worried that an unstoppable tide of economics imperialism might render their subject specific expert knowledge increasingly redundant. Such fears of disciplinary displacement are only too real, and they will certainly persist whatever the reaction to the ideas in this book. The trend is too well established for it to simply disappear, however well written the guide to how best to mobilize resistance. And *False Prophets of Economics Imperialism* should not be mistaken for such a guide. The outer reach of its aims is to show that economists' market models are not a direct substitute for the empirical knowledge of subject specialists. The former speaks to self-made phenomenal worlds in which purely abstract relationships stand in for actual human experiences, and the latter speaks to something that we might recognize directly from everyday experience. One style of argument should not be expected to displace the other, because they exist on separate ontological planes. I would therefore urge greater boldness amongst social scientific subject specialists in emphasizing the distinctiveness of their research. There are lots of things that it can do that the mathematical market models of the economics imperialists simply cannot. Much of the shoutiness of existing debates revolves around attempts to settle disputes about which is superior, but a more straightforward recognition of difference should be sufficient to heighten the confidence of the subject specialists that their role in the future remains secure.

The philosophical chapters highlight the difficulty of overcoming the separation between the world as depicted by a mathematical market model and the world as it is experienced in everyday life. Economic modellers have largely given up trying to draw direct inferences from their self-made phenomenal worlds to the world beyond, but they are still likely to claim epistemic functions on behalf of the mathematical market models they introduce into other social sciences. However, the historical chapters show that the foundations on which such attempts build are far from assured. I study four periods in the prehistory of economics imperialism, the latest of which concluded 20 years before anyone started to use the phrase as a positive marker of association. My analysis focuses on the work of Stanley Jevons in the 1870s, Lionel Robbins in the 1930s, Paul Samuelson in the 1940s, and Kenneth Arrow and

Gérard Debreu in the 1950s. Viewed sequentially, what is most noticeable is how much more sophisticated the mathematical instincts became in each successive phase, even as the introduction of ever greater abstract mathematical content came at the expense of substantive economic content. Market models displayed the imprint of mathematization in each of these phases, but differently so. This meant that each of the pioneers wrestled with the question of how much the structure of lived economic relationships had to be represented within their market models, even if the underlying mathematical mode of reasoning ensured that the answer was never fully convincing philosophically. By the time of Arrow and Debreu, those who were operating at the furthest frontiers of mathematical economics had decided that the model world should be allowed to stand on its own, its truth claims unencumbered by restrictions imposed upon them by everyday life. Even then, it was possible to demonstrate as a matter of mathematical logic that market models did not work in the way that had always been assumed and, therefore, their truth claims were subjected to strict ontological constraints.

Again, this is not intended as a wholesale dismissal of economists' use of mathematical market models. It is only a plea to recognize what sort of arguments their use will licence and what it will not. Economic modellers have largely made peace with the idea that the epistemic capacity contained within their mathematical objects is not as straightforward as direct inference. Social scientists fearful that their specialist literatures are in the process of being overrun by economics imperialists should take heart. The arguments that are permitted by the introduction of mathematical market models are of a completely different nature to the arguments of the subject specialists. This is because economics imperialists must begin by treating even fundamentally non-market social experiences as if they were unproblematic market phenomena. Subject specialists are better placed because of their usual commitment to contextual empirical knowledge to treat social experiences as they actually are. Economics imperialists can reveal the underlying logic of action in circumstances in which all human behaviour responds to nothing other than a pure market frame of reference, whereas subject specialists are always likely to search for deeper, more complex and less predictable drivers of real-world behaviour. Anyone who tries to say that these are such sufficiently similar objectives that ultimately only one can survive has seriously underestimated the essential difference between the world within the model and the world beyond. There will always be room for detailed subject specialist knowledge. If there is a long-term threat to the social sciences, it lies in downgrading the need to capture the nuance, the subtlety and the essential contingency of constantly evolving social formations in favour of one-size-fits-all explanations.

I have so many people to thank for helping to shape the thinking underpinning this book that it would be impossible to name them all individually. This is always likely to be the case for a book that has had as long a gestation period as this one. I have been fascinated by attempts to stipulate the limits of mathematical market models since my days as an undergraduate economics student in the early 1990s. The start of my ESRC Professorial Fellowship (www.warwick.ac.uk/rethinkingthemarket) therefore came 20 years into the process of reflecting on the essential differences between model-based knowledge and empirical knowledge. My earliest lecturers encouraged me to think aloud on these matters even when I had little other than intuition to go on, and they could easily have used their vastly superior learning to stamp on what I am sure they often took to be my rather unhelpful convictions. I subsequently spoke at greater length on related issues with many more people when I was completing my PhD and then beginning to make my way as an early career scholar. If they were to say it was in any way obvious that my thinking back then would develop into what can be found in the pages that follow, this merely shows that they had more faith in my abilities than I did. I would like to thank them all for being so generous with their time and for treating with good humour my attempts to repeatedly turn the conversation back to mathematical market models, whatever it had started out as. I also wish to thank more recent colleagues for further inspiring chats. You have all left a larger imprint on this book than I am sure you would ever have thought possible.

Thanks also to my publishers at Agenda, Alison Howson and Steven Gerrard. Once again, they have given me my head to make the argument in the way that made most sense to me. I am very grateful for that and look forward to working with them again on my next endeavour.

As ever, I have saved my most important thanks until last. They are, of course, to Katie. We were required to shield her from Covid for three-and-a-bit years, from March 2020 to May 2023. The cats thought this was wonderful, as they had two people around all the time to cater to their every whim. But it was more challenging for one of their humans than the other. As our long vigil of social isolation coincided with me completing most of the writing tasks for

False Prophets of Economics Imperialism, I always had work to throw myself into to while away the time. However, when Katie was restricted to having to rely on me for her whole human contact, she would often find me distracted by the finer philosophical and historical points of hypothetical mathematical modelling. I offer her my apologies and thanks in equal measure. This book is my latest to be dedicated to Katie, but it will not be the last.

Matthew Watson

Introduction: economics imperialism and social science

THE REVOLUTIONARY INTENT IN ECONOMICS IMPERIALISM

This book is about methodological revolutionaries. It discusses a first group of people who changed fundamentally what economists accepted as good theoretical practice. I show how these intellectual pioneers reduced the most essential elements of economic theory to working through the logical implications of beginning with a market model and subjecting it to modes of mathematical reasoning. I also discuss a second group of people who changed equally fundamentally economists' willingness to position their mathematical market models as the one true means of generating valid social explanations. I show how these later trailblazers made the case for an economics imperialism that displays little respect for the established domains of the individual social sciences. The earlier reconstitution of economic theory in ever more noticeably mathematical form can be thought of as the prehistory of today's increasingly prominent boundary-hopping activities. None of the former group ever made the case explicitly for economics imperialism, but without their theoretical innovations the scope for subsequent disciplinary transgressions would have been far more circumscribed. More than a century of deep methodological reflection *within* economics about how best to mathematize market models ultimately created the context for a debate about how far the resulting mathematical objects could be taken *beyond* economics.

These revolutionary activities have led to a situation in which all manner of non-market aspects of everyday life come to be treated for analytical purposes

1

as if they are market phenomena. However, this requires not only a change of explanatory focus but also a reconceptualization of the social realities being studied (Kuorikoski & Lehtinen 2010: 357). The latter shift is clearly the more controversial, because it involves accepting that something is not as it is experienced in real life for the sake of methodological convenience. The use of market models to explain various aspects of social life amounts to more than mere deference to the most up-to-date techniques. It tells us how we must view what is of interest in the world, even if we struggle to be convinced that every element of our daily lives can be reduced to simple market calculations. Of the many resulting ironies, perhaps one stands out above all others. My first group of revolutionaries made economic theory the province of mathematical reasoning tools. Argument through mathematical analogy increasingly crowded out verifiable economic content from economic theory, leaving behind a noticeably uneconomic economics (Watson 2018: 20). Yet the same mathematical reasoning tools were then used by my second group of revolutionaries to colonize other social scientific subject matter. In the process it forced myriad social phenomena to be imagined as if they had acquired the most basic economic characteristics, even as economics itself became increasingly uneconomic.

For instance, the law and economics literature has attempted to reconfigure how to understand developments in legal practice (e.g., Shavell 2004; Cooter & Ulen 2013; Butler & Klick 2018). Where once it had been standard to think of them through the non-market lens of the interaction between legal precedent and the changing structure of public morals, now the same questions are also considered through the market lens of maximally efficient legal statutes (e.g., Becker 1968; Landes & Posner 1993; Cooter 2000; Zerbe 2001; Parisi 2004). The same shift has occurred in the political science literature on party competition (e.g., Grofman 1993; Hinich & Munger 1997; Dunleavy 2013). Where once a party's ideology was considered to provide it with a series of specific political messages to convey to the electorate, now party competition is as likely to be presented in terms of the purely functional task of identifying the electoral market equilibrium reflected in the wishes of the median voter (e.g., Riker 1962; Davis, Hinich & Ordeshook 1970; Stigler 1972; Wittman 1973; Robertson 1976). Likewise the sociological literature on labour market discrimination (e.g., Lundahl & Wadensjö 1984; Rodgers 2006; Donohue 2013). Where once the issue was approached through the lack of legal safeguards against unequal treatment in a society in which discriminatory attitudes ran deep, now there are counterexplanations focused on market rewards for incentivizing non-discriminatory behaviour (e.g., Arrow 1972; Phelps 1972; Akerlof 1976; Neumark 1988; Oaxaca & Ransom 1994). Similarly with the climate science literature, where mathematical market models have been used to challenge the very conception of environmental sustainability based on empirical studies of the effects on nature of modern systems of production (e.g., Dornbusch & Poterba 1991; Schwarze 2001; Mendelsohn 2022). Climate scientists had long argued that the earth has a finite carrying capacity which can easily be exceeded when seeking to satisfy untamed economic desires. Economists armed with market models subsequently said that this was to miss the point because these problems would disappear under an efficient allocation of property rights (e.g., Ostrom 1990; Nordhaus 1991, 1994; Frankhauser 1995; Tol 2009; Eboli, Parrado & Roson 2010). Similarly too with the economic history literature, where mathematical market models have been used to overturn consensus opinions built on deep archival research of what the actors involved thought at the time (e.g., Whaples 1991; Lyons, Cain & Williamson 2008; Greasley & Oxley 2011; Diebolt & Haupert 2015). Historians had long argued, for example, that the American civil war was fought to secure the purely moral objective of ending the system of slavery because that system was in any case economically inefficient, as well as that the course of twentieth-century US economic development was conditional upon the prior construction of a continent-wide railroad system. Economists subsequently used market models to tell the historians they were wrong: from their market perspective slavery was to be seen as a rational system of production but building railroads an irrational infrastructure investment (e.g., Conrad & Meyer 1958; North 1961; Fogel 1964; Fogel & Engerman 1974; Drescher & Engerman 1998).

In all of these instances a shift has occurred from a non-market lens to a market lens, simply to make space for the insertion of mathematical reasoning tools. These are illustrations of economics imperialism in action, and many more besides could have been used in their place. Until now, though, almost none of the critical commentary on such developments has focused specifically on the mathematical sources of the change in interpretive lens. It is already well established that substituting a market lens for a non-market lens has increasingly destabilized the border markers between economics and the other social sciences. But there has yet to be a sustained analysis of how these ever more porous boundaries result from increasingly handing over economic theory to mathematical analogy.

These changes also have clear policy consequences, because each of the above examples comes with its own in-built policy solution. Whatever questions lawmakers are asking, the answer is always to establish a market structure that allows individuals to select the best available option. However, studying the policy consequences of economics imperialism is another book for another day. It is more than enough for now to chart the intellectual route that allowed economics imperialism to gain a foothold in various social scientific literatures. After all, economics imperialism currently tends to be treated as an action to be performed rather than a professional state of mind whose underlying intellectual rationale needs to be defended. This book reverses such polarities.

It has been written primarily for those who are worried that their subject specialist knowledge now counts for less in the face of the apparently indiscriminate wielding of mathematical market models. There is understandable anxiety that the premium that was once placed on knowing a particular aspect of the social world in often minute detail will disappear in the face of economics imperialists' one-size-fits-all method. Subject specialists tend to advocate an approach that ties explanation securely to the known operation of the real phenomena in which they are interested. Economics imperialists embed the alternative standard that the key to good social science lies in adopting a rigorous method for stipulating the relationships that enter mathematical market models, not in attention to detailed investigations of lived experiences in the world beyond. Their goal is to rigour-check rather than fact-check the existing state of knowledge on topics where they are often unabashed about signalling their own empirical ignorance. The anxiety of subject specialists results from the fact that knowing more than their opponents may no longer be enough. At heart, this is a clash over different ways to make arguments, but one that will play out increasingly unequally the more that the economics imperialists are able to assert their standards of mathematical competence as a universal objective.

The same move towards rigour-checking had already been made in economic theory many years before, when it first became conventional to define economics as a way of thinking rather than in relation to its own particular subject matter. This was a reorientation away from increasing the stock of what might be known to drawing strict lines between what is worth knowing and what is not. Viewed from this perspective, economists should not limit themselves contents-wise to what they are most familiar with, as long as in adopting their mathematical market models they are following the appropriate method.

The key publication that encouraged them to think more expansively was Lionel Robbins's 1932 book, *An Essay on the Nature and Significance of Economic Science*. Robbins (1932: xlii) noted that economists were already following dual tracks: a newer generation was branching out to treat economic theory as a means of studying abstract choice situations; their predecessors were still wedded to trying to understand the dynamics of material wellbeing in a well-governed society. Robbins threw his lot in with the insurgents and subsequently changed economics for good. His self-professed aim was to create an autonomous sphere in which economic theory might operate, stripped bare of other legal, social, cultural and political influences. Vilfredo Pareto (1999 [1900]: 255) had previously observed that economics might prosper as a distinct discipline if it were to be recast as a science of choice. "I do not have the slightest difficulty", he said, "in justifying the assertion that pure economics is concerned only with the choices which fall on things the quantities of which are variable and susceptible of measurement". Robbins (1984 [1935]: 16) then used this as the basis for an analytical definition through which economics could be understood as the study of competing alternatives under conditions of scarcity. His accompanying dismissal of what he called classificatory definitions severed the links between economics and "the economy" and established the context for the border skirmishes that followed.

THE IMPRINTS OF MATHEMATICAL LOGIC IN ECONOMICS IMPERIALISM

Judging by the research that appears in the discipline's leading journals, economists seem content to assume that the path to professional prestige requires them to spell out the content of their mathematical models from scratch

(Backhouse 2012: 42). It often appears that the greatest reputational gains arise from the aesthetic elegance of the solution to the system of equations, not from what can be learnt about the economy from reasoning in this way (Shiller 2012: 132). However, when comparing this to the work published by economics imperialists in other social science disciplines, important differences emerge. The look of the respective research is very different for a start. Economics imperialists will still on occasions lead the reader through their mathematical market models from scratch, but this tends to be the exception rather than the rule. It is usually reserved for economics imperialists' own in-house journals, but when they publish in other disciplines' journals it is likely to be in a form that is more obviously mathematics-lite. This does not mean it is anything other than a mathematical market model being used as the reasoning tool, only that it is shorn of much of its mathematical finery. It is a sign of how commonplace this way of thinking now is within the social sciences that it is no longer necessary to build the model from scratch on every occasion. The ability to spot the readily recognisable imprints that the mathematics leaves behind is enough.

Those features are easily listed. They include an abstract agent cast very much in the image of the hypothetical *Homo economicus*; a form of agential rationality that removes the individual from the broader social context that gives more rounded meaning to their life; a decision-making matrix where outcomes can be ranked from the most to the least efficient; and an explanation of action that requires all decisions to be perfectly aligned with what is best for the individual. To apply mathematical market models to myriad social situations is therefore to place a hypothetical market agent at the centre of the analysis. It is for this reason that I focus so much attention on the progressively tighter restrictions that were placed on the conception of the acting individual in each phase of the prehistory of economics imperialism (see Chapters 4–7). It experienced a series of reductions, to the point where it stopped being asked whether what remained resembled a real person in any way, because it was obvious that it did not.

Economics imperialists in the Robbins mould thus use the image of an instrumentally rational *Homo economicus* to justify turning all situations of choice specifically into situations of market choice. Given the significance of this shift, it is worth repeating that this is not only to change the character of the *explanans* (that doing the explaining), but also the character of the

6

explanandum (that being explained). There is no need on each occasion to present a formal derivation of market demand and market supply schedules, nor yet of the indifference and isoquant curves that underpin them. All that is required is to place a hypothetical market agent in a correspondingly hypothetical market setting. Everything that lies behind these conceptual configurations can be taken as given. However, this predetermines the nature of the choices available so that the only one it is possible to make is that which leads to the mathematical solution of the system of equations. Once the character of the choice environment has been recast as a specifically market setting, then the market must be made to clear at equilibrium, otherwise all the preceding restrictions on agential behaviour will have been for nought.

Such supremely rational choices disqualify any sense of the agent feeling that they are in charge of their own destiny (Hay 2002: 103). In Martin Hollis's (1998: 16) words, they are "the very model of a modern individual". But they attain this status only on the back of an unwritten pact through which everyone allows their decisions to be made for them, so that the needs of the equilibrium solution impose themselves on all action. The mathematics of market models exercise full control over what type of behaviour is allowed (Blakely 2020: 43). An inversion thus occurs, whereby it is no longer consciously acting individuals whose decisions scale up to an equilibrium solution. It was a prized goal of the late nineteenth-century marginalist revolution to prove how the aggregation process worked as an *economic* matter, but orthodox economic theory has long since retreated from this particular objective (see Chapter 7). Instead, the equilibrium solution strips individuals of conscious action and makes their decisions for them.

This acts as a mechanism for enabling the mathematical logic to operate as intended. Even when the mathematical content drops out of the presentation of the argument, the underlying mathematical logic still has to be protected from the misbehaviour of individuals who insist on thinking for themselves. I return repeatedly in future chapters to the question of how the economic agent was increasingly purged of its connection to the real world. But I do so specifically to show what it was in the surrounding mathematical logic that required such a move (see Chapters 4–7). The prehistory of economics imperialism is the story of accepting an increasingly uneconomic economics as the price for introducing ever more precise mathematical applications into market models.

However, by prioritizing the mathematics over the economics of the equilibrium concept, the locus of the explanation also changes. In no known state of the world do people simply act out the solution concepts of a system of equations. But the mathematical market models that economics imperialists use to cross disciplinary boundaries do not rely on real people to populate their explanatory schema. To use Mary Morgan's (2012: 30) terminology, their explanations relate to actions that take place in the self-made world within the model, not to the messier real-world dynamics that we can be forgiven for assuming the model is intended to represent. The difference between the two will continue to be emphasized in what follows, as it captures a key distinction between the research of the economics imperialists and the research of the subject specialists they are seeking to displace. It also tells us about the explanatory limits of mathematical market models. Most social science research is animated by a concern to reveal the characteristics of lives that are actually being lived. Yet economics imperialists most obviously focus only on the implications of building their models through particular abstractions, and the world being explained is consequently that within the model.

TIMELINE TO ECONOMICS IMPERIALISM

It would be expecting too much to try to turn the following account into a broader history of forgone alternatives within economic theory. There were countless paths opened up between the 1870s and the 1970s, and looking back today most of these ultimately appear to be roads not taken. But it is impossible to attempt a history of everything, and that is why I have set my sights more narrowly. My interest is those moments in the history of economic thought that ushered in new forms of theoretical endeavour, but only where the resulting changes seem to have been crucial to the prehistory of economics imperialism. In particular, I emphasize four distinct confrontations between market-based economic theory and advances in mathematical insight: the marginalism of Stanley Jevons (Chapter 4); the economization of Lionel Robbins (Chapter 5); the maximization of Paul Samuelson (Chapter 6); and the axiomatization of Kenneth Arrow and Gérard Debreu (Chapter 7). Each of these constitutes both a revolutionary moment in economic theory and a foundation stone for economics imperialists' future revolutionary actions.

It is important, though, to keep these cycles analytically distinct, because they were set against different historical backdrops. The pioneers of new forms of mathematical market models gave no thought to the economics imperialism of the future. Their sights were set elsewhere, most obviously in trying to fend off the doubts that arose in their own time about their vision of what economics could be. In the 1870s, Jevons had to position himself not only against the increasingly sterile Millian classical economics of the day, but also a well-established tradition of historical economics and nascent traditions of institutionalist and Marxist economics. In the unprecedentedly dynamic decade of the 1930s – one that George Shackle (1967) dubbed the "years of high theory" – Robbins was only one voice amongst very many arguing for different approaches to the most foundational questions of economic theory. In the 1940s, Samuelson was faced with making a first-principles case for the rejuvenation of classical microeconomic theory when all around him others were working backwards from Keynesian macroeconomics towards its accompanying microfoundations. In the 1950s, Arrow and Debreu were operating against an established Samuelsonian orthodoxy while seeking to popularize forms of mathematical expression that were well beyond the comprehension of the vast majority of their fellow economists. They were all doing their own thing in their own way, within a context in which nobody was talking explicitly about economics imperialism. My decision to understand Jevons, Robbins, Samuelson, Arrow and Debreu in relation to the prehistory of economics imperialism is therefore a *post hoc* reconstruction of my own making.

I have been careful to locate them within the debates about mathematical economics that they were having, rather than jumping to the whiggish conclusion that their solutions pointed inevitably to what came next. It will hopefully be obvious when reading Chapters 4–7 that the exact opposite was more often the case. Still, though, whenever faced with a *post hoc* reconstruction, there might be reasons to object to the choice of characters to base the historical narrative around. It could be said that I have identified the wrong revolutionaries, with the theoretical innovations I attribute to Jevons, Robbins, Samuelson, Arrow and Debreu belonging more properly to someone else. Other perhaps lesser-known economists might have plausible claims to priority on purely theoretical matters – von Thünen or Gossen for Jevons, von Wieser or Wicksteed for Robbins, Cassel or Hicks for Samuelson, Wald or

McKenzie for Arrow and Debreu – but my selection looks sound when the developments of interest are simultaneously theoretical and mathematical.

As will become clear, when thinking about economics imperialism it is not just advances in economic theory that matter. It is how developments in economics, mathematics and philosophy combine with one another. To make it easier to visualize what was going on when, I have constructed a timeline to show the time at which the most important arguments for my appraisal of economics imperialism first came to their respective literatures (see the Timeline, pp. 243–6). When set out in such a way, it is immediately apparent how late the discussion about economics imperialism started relative to the developments in economic theory on which it is based. My prehistory must therefore be viewed as literally that, something which was complete long before the advent of widespread advocacy for economics imperialism. It is also noticeable that there were a number of arguments already in circulation which, had they been more widely known, might have been expected to have derailed the economics imperialism bandwagon before it had gained any momentum. This was true of both the mathematics and philosophy literatures. It had already been shown there that explanations based solely on mathematical logic, interesting as they were as thought experiments, had no means of demonstrating that they were more than a multipurpose template overlaying what were, in essence, wholly dissimilar phenomena. More suggestively, this was also true of the economics literature. The first tub-thumping account of why economists should feel emboldened to take their mathematical market models anywhere they pleased was published in 1984 (Stigler 1984). Yet this was *after* the announcement of the most dramatic finding of the axiomatic phase of mathematical economics: that the basic market model was unfit for purpose even when attempting to explain unequivocally market outcomes (Hildenbrand 1983). The chronology of the economic, mathematical and philosophical arguments underpinning my appraisal of economics imperialism is therefore nothing if not convoluted.

STRUCTURE OF THE BOOK

The remaining chapters tell the story of how economic theory came to focus increasingly on the mathematical features of the world within an explicitly market model. The scope for economics imperialism was much enlarged as a consequence, requiring only the reinterpretation of other aspects of social existence in line with the solution concepts that determine the division between a market world and a non-market world. However, such moves provide no grounds for convincing anyone to believe that social relations of lived experience actually mirror the only ones the model world can describe. This is not intended as a criticism of the use of mathematical market models, only an acknowledgement of limits. Economics imperialists have failed to show that the social world is as unified in nature as it needs to be if the introduction of market models across disciplinary borders is to lose its current rather indiscriminate feel. The unification of social scientific theories under the umbrella explanation provided by mathematical market models is not in itself justification for those theories. Only the generation of more plausible explanations of actual lived experiences would allow them to pass such a test (see Chapter 1).

However, economists have long since embraced the demands of formalisms that require the explanation to be situated within the mathematics of their models (Backhouse 2012: 44). This poses the two fundamental questions around which the rest of the book is organized: how and when did this shift occur? The "how" question relates to the process through which the closed-system dynamics of mathematical market models became an acceptable substitute for the open nature of real social systems. It must be addressed philosophically and it provides the focus of Chapters 1–3. The "when" question relates to the timing of the exclusion of substantive content from the world within economic models. It must be addressed historically and provides the focus of Chapters 4–7. My objective is to show that nothing in the claims made today in the name of economics imperialism was in any sense preordained in the developments of economic theory which came to serve as important staging posts in its prehistory. Economics imperialism is therefore a much more conditional and a much more contingent process than its proponents like to make out when referring to it as the inevitable pinnacle of social science research.

Chapter 1 provides the necessary background for all that comes next. Without ever doing so explicitly, the most vocal proponents of economics imperialism appeal to the image of scientific unification to make their case for using mathematical market models for explaining non-market social phenomena. A unified social science is better than the coexistence of disparate disciplines, so the argument goes, and best of all is unification through subordination to mathematical market models. Many years ago, though, the science history and philosophy of science literatures moved past the general proposition that scientific unification is always a good thing. Historians of science have shown how often the stand-out moments of scientific unification have involved bringing together different physical phenomena under a common mathematical framework. This is obviously not the same as showing that those phenomena have an identical essence in practice. Philosophers of science now often distinguish as a consequence between derivational and ontological unification. The redefinition of non-market social phenomena to facilitate the imposition of mathematical market models does nothing to demonstrate that all social phenomena thus explained are ontologically alike. Philosophically speaking, then, economics imperialism would seem to be based on derivational unification.

Chapter 2 investigates the origins of allowing the application of common mathematical methods to take precedence over detailed empirical explanations of phenomena as they actually are. It focuses on turn-of-the-twentieth century mathematician David Hilbert's attempts to reconstruct what was involved in making a mathematical argument. These were not entirely successful attempts, but they have left an important mark on the prehistory of economics imperialism. Mathematical proofs were brought to life pre-Hilbert through demonstration effects. This could be through either systematic observation of real physical phenomena or the painstaking task of calculating the relevant number chains to the point at which the future discovery of a disconfirmatory sequence could be safely ruled out. Hilbert, by contrast, began with a model, and he demanded only that it obeyed the existing laws of logic. This was enough to make it true of itself, and there was no need for additional external validation of the truth claims that are made solely in relation to the world within the model. Hilbert changed the very notion of what it meant to construct a rigorous mathematical model. He located the practice of proof-making in a different sphere to that of tests involving real number sequences, consequently removing all ontological requirements from mathematical postulates.

Chapter 3 moves the discussion from explanation through mathematical postulation to explanation through hypothetical mathematical modelling. These are the two key trends in the mathematization of economists' market models. However, hypothetical mathematical modelling in economics tends to move somewhat interchangeably between described mathematical functions (where the solution has actually been found) and defined mathematical functions (where it is merely logically findable). Aspects from both sides of this ostensibly binary divide routinely appear in the same model. Economists typically suggest that their models have inferential capacity to the real world (as per described mathematical functions), but without expressly laying out why this might be (as per defined mathematical functions). To its advocates, economics imperialism is the future for all of social science because it provides the analytical conditions for enhanced inferential potency. Yet, at most, the development of mathematical structures as standalone artefacts embedded in generally accepted principles of economic theory allows only for inferential statements to be confirmed by the theory. It sidesteps the need for external validation via real-world observations. In imposing their own mathematical market models onto non-market social phenomena, economics imperialists are thus restricting the knowledge that can be produced to that which is consistent with an abstract account of a predetermined equilibrium solution.

It is always unwise to think that economists working many years ago would have had access to the language used in philosophical debates today. The connection between Chapters 1–3 and Chapters 4–7 is much less direct than this. But some careful retrospective analysis can alert us to decisive turning points in the relationship between economic theory and mathematical method, which can then be reflected back through contemporary debates in both science history and the philosophy of science. This assists in drawing attention to the all-important moments when the economic content of market models was supplanted by mathematical content, with the goal of ontological unification accordingly giving way to that of derivational unification.

Chapter 4 studies the mathematical innovations brought to economists' market models by the work of Stanley Jevons, particularly his 1871 *Theory of Political Economy*. By thinking in terms of decisions occurring at the margin – the last possible decisions of that nature to conform to rational responses to price signals – marginalism paved the way for the development of theories of equilibrium. From there, it became possible to conceive of every decision across all social domains as if they were intrinsically the same, as long as they could be modelled specifically as a market decision. Yet Jevons did not make such a large leap. To his mind, mathematization of the market model really meant the enumeration of every possible feature of market life. Jevons used the differential calculus as a mode of thinking, but not solely as a means of logical expression of immanent possibilities. He aimed always for fully described mathematical functions based on discovery through experimentation of the real value of numbers appearing in his systems of equations. Jevons's mathematical market models were therefore not merely mental constructs. He demanded that they act as a direct representation of the world beyond the model, making present known aspects of the real world within his market models.

Chapter 5 discusses the contribution made by Lionel Robbins's scarcity definition to promoting the concept of economization. Jevons had attempted to separate economics from the other social sciences on the basis of subject matter, but Robbins's 1932 *Essay on the Nature and Significance of Economic Science* did so by isolating a single aspect of behaviour. He believed that unless people were making the very most out of the scarce means at their disposal, it could not be said that they were acting economically in any formal sense. It was how they chose to act that mattered to economists, not the domain in which those actions took place. By reducing economics to the study of economizing conduct, Robbins provided the discipline with a more tightly circumscribed axiom base, which in turn revealed rich pickings for hypothetical mathematical modelling. However, this was not a potential Robbins sought to exploit himself. He is the one great exception in the historical part of the book, because his instincts were as anti-mathematical as anyone who was working in the theoretically eclectic 1930s. His vision for economics was fully immersed in mathematical logic, but his scepticism only increased over time when it came to handing over economic theory to mathematical content.

Chapter 6 shows how Paul Samuelson's 1947 *Foundations of Economic Analysis* positively knocked down the mathematical door that Robbins had left ajar. He took Robbins's reduction of economic theory to the study of choice situations and turned it into the logical working through of one situation after another as intrinsically the same maximization problem. Samuelson flirted with the potential unification of the social sciences by first seeking to unify economics through the application many times over of what was, at heart, the same solution to the same system of equations. He simply moved from one economic topic to the next, armed with inherently the same mathematical reasoning tool. In the process, he changed fundamentally what it was necessary to know to demonstrate familiarity with economic theory and, accordingly, he also changed the whole *look* of economics. Number ceded priority to symbolic notation, as real phenomena were replaced by the logical delineation of abstract conditions of existence. The scientific unification in evidence here is derivational rather than ontological, marking a clear break with both Jevons and Robbins. There is no requirement for the underlying real phenomena to be essentially the same, because no degree of difference will prevent their abstract analogical selves from behaving in identical fashion within formally identical mathematical models.

Chapter 7 rounds off the historical study by exploring how Kenneth Arrow and Gérard Debreu brought the technique of formal axiomatization to the mathematization of market models. Their famous 1954 *Econometrica* article contained the first complete existence proof for the conditions of general equilibrium, in doing so introducing pure Hilbertian mathematical postulation into economics. Under such a structure, hypothetical mathematical models are true simply on the grounds of having had the correct logical procedures applied to them. These are evidently defined rather than described mathematical functions, and they obey no ontological constraints arising from known facts about real phenomena. Arrow was nonetheless always searching for economically interpretable practical applications of his mathematical proofs, even if his major successes in this regard came only in highlighting how limited market theory was in capturing the essence of real-life situations. Meanwhile, Debreu insisted that his systems of equations solved purely mathematical problems that were entirely separate from accompanying attempts to infer economic knowledge. Either way, the world within the model is solely an artefact of assuming that economic theory reaches a heightened state of rigour when it is embedded within mathematical functions that can be considered true in an axiomatic sense, without the need to demonstrate any obvious connections to reality.

The Conclusion shows how far the prehistory of economics imperialism travelled in only 80 years from Jevons's attempts to enumerate elements of lived market realities to Arrow and Debreu's existence proof of an infinitely systematized market model. However, the comforting stories that the proponents of economics imperialism tell themselves about their approach's predestined superiority finds no support under sustained historical treatment. In each of the marginalism, economization, maximization and axiomatization phases, the trailblazers remained genuinely unsure about just what they had achieved. Jevons wanted his enumeration strategy to leave fundamentally intact his economic agent as a recognisable person from everyday life, but he did not know what to do about the problem of free will (see Chapter 4). Robbins had a clear vision for restricting the scope of economic theory, but he readily acknowledged that his colleagues would then have nothing of obvious value to say about the real world (see Chapter 5). Samuelson assumed that his new techniques would rid economists of the need to stipulate why their hypothetical market agents act as they do, but he discovered that his mathematical models only reached their predetermined solutions by asserting some such behavioural propensity (see Chapter 6). As the supreme mathematical economists, Arrow and Debreu were in the best position to turn market models into purely mathematical artefacts, but they found that there were no economically justifiable assumptions on which to base their mathematical market models (see Chapter 7).

What, then, are these findings likely to do for those who are situated on competing sides of the economics imperialism debate? It is not to be expected that it will change the minds of anyone who has already committed themselves to the colonizing cause. They almost certainly already have too much invested to consider writing off substantial career sunk costs. As such, those who are anxious about their own subject field's ability to withstand challenges from mathematical market models are unlikely to have their concerns assuaged. The following chapters hold only meagre potential for quelling other social scientists' *professional* fears about the future. *Intellectual* fears, however, are a different matter. The remainder of the book reveals the largely baseless nature of economics imperialists' trumpeting of their own superiority. To say merely that such claims are built on shaky foundations is to be extremely generous to them. They should hold no purely intellectual terrors for other social scientists whose commitment to detailed studies of empirical reality continues to provide their work with an important point of difference.

CHAPTER 1

Setting the scene: scientific unification through the use of mathematical models

INTRODUCTION

The case for economics imperialism is a specific example of the broader argument for scientific unification. But it is not a matter of using similar means of observation and measurement to adjudicate on the similarity between the causal processes in operation. The explanation does not follow careful empirical investigation of the causes at work; rather, it comes first. Every social situation is reduced to an instrumentally rational individual ignoring conflicting social stimuli to always maximize their market gains, and an explanatory narrative is then built around those actions. Explanatory notions are thus allowed to overpower causal notions, and scientific unification very quickly becomes the search for new social situations to submit to a single structure of explanation (Kitcher 1989: 495). The boundary-hopping activities in which economics imperialists engage are one-way transgressions enacted via the imposition of mathematical market models (Nik-Khah & Van Horn 2015: 72).

These are circumstances in which no invitation to appropriate new subject matter is ever sought, and what might otherwise be a negotiated coming together instead gives way to the law of the jungle. Economics imperialists are easy to convince that their side will be the last one standing in any ensuing battle for supremacy (Van Bouwel 2011: 47). Listening only to them, the sense of an imminent takeover is never far away (Fine 2004: 121). Looking at the practice in social science journals, the same feeling can also quite easily take hold. It is one of the curiosities of economics imperialism that very few members of its club make the case explicitly for it. But for every economist who has spoken loudly in support of transgressive mathematical market models, there are many more who have followed their prescriptions but without the fanfare. Even if they are concerned not to openly ruffle feathers, they are equally intent on pushing the parameters of insisting on strict obedience to an abstract market logic.

For such an important trend in social scientific explanation, the historical background to contemporary economics imperialism is surprisingly murky. Even the history of the term seems to invite two contrasting narratives. The most vocal proponents of economics imperialism have tended to coalesce around an important 1968 speech delivered by Kenneth Boulding as the moment the term was first used. It is difficult to miss the hostility in Boulding's (1969: 8) analysis of his colleagues' then-recent self-insertions into neighbouring intellectual territories. Yet card-carrying economics imperialists seem to have treated it as a badge of honour that Boulding would have used a speaking engagement as prestigious as a presidential address to the American Economic Association to warn them to go no further. It is evident that they consider themselves to be rattlers of cages, upsetters of apple carts, challengers of convention, and therefore if they had managed to offend an establishment figure quite so badly then they must have been doing something right. They have always worn proudly the disruptor's desire to appropriate as a marker of belonging a phrase that was initially used pejoratively against them.

However, it seems to have escaped their attention that the same phrase was first used 35 years previously to forward a completely different argument. In 1933, Ralph Souter (1933a: 377) took Lionel Robbins's *Essay* to task for emasculating economic theory, subjecting it to "[s]uffocation through analytical definition" (see Introduction). It was in this context that he wrote about "economic imperialism" (without the "s" that is more commonly used today) (Souter 1933b: 94). Robbins's reconstitution of the subject field as the study of abstract choice problems under conditions of scarcity changed fundamentally the notion of precision to which economists' work would be aligned (Medema 1997: 123). Souter (1933a: 378) was concerned that all which might be left was purely "logical precision", in which the solution to economic problems was contained in the working through of a model's initial mathematical premises. He much preferred Alfred Marshall's conception of precision, which required matching theory with reality rather than deriving theoretical statements axi-omatically. Initial abstractions – what Marshall (2013 [1890/1920]: 36) called the essential acts of "imagination" allowing economists to be put "on the track of those causes of visible events which are remote or lie below the surface" – could be rendered ever more concrete. This is consistent with what philosophers have subsequently called the "de-idealization" of the assumptions on which hypothetical mathematical models are constructed (McMullin 1985: 261; see Chapter 3). Achieving precision in this instance follows from diligent empirical study of the institutional basis of everyday economic life and of the effects of legal, social, cultural and political institutions on contextually specific enactments of individual economic agency. Souter (1933a: 378, 377) mourned the eclipse of Marshallian precision at Robbins's hands, equating the resulting "logical precision" with "static formalism".

Souter's (1933b: 94) attempted fightback involved advising his colleagues that they should "invade the territories of [their] neighbours" in the search for the detailed empirical knowledge that the Robbins definition would banish from economics. His notion of an invasion is very different, though, to what it means today to the most forthright economics imperialists. Souter's vision for social science was one where the boundary hopping of economists would be designed to enrich their own subject field, not to displace others from what they considered their natural home. His transgressions would be reciprocal, requested interactions operating as two-way crossings. It is consequently unsurprising that modern-day proponents of economics imperialism have been selective in their memories, choosing a timeline for the phrase that suits their own contemporary self-image. It saves them from having to explain what happened to the alternative future that never was, economics imperialism in the form of the neighbourly non-aggression pact envisioned by Souter.

Two models of scientific unification are thus being invoked. Souter (1933b: 95–7) initially wrote of a purely consensual realignment of the social sciences that would build in cumulative fashion on what was best across all of them. This is scientific unification through collaborative accumulation of empirical facts across multiple domains. Explanation is attempted only once the causal processes have been established empirically and only once as much conceptual de-idealization as is possible has been achieved. By contrast, the economics imperialism that Boulding (1969: 7–10) lamented put the explanatory content first and subordinated the causal content to it. Conceptual de-idealization did not get a look in, because the mathematical assumptions on which Robbins's logical precision rested allow no role for it. For the equilibrium solution of the market model to assert itself, the mathematical assumptions must be left to stand on their own, relieved of too much concern for empirical realities.

In an attempt to further distinguish Souter's notion of economics imperialism from the road eventually taken, the chapter now proceeds in three stages. In section one, I draw attention to the dating of the most important claims made in the name of economics imperialism (see also the Timeline). These occurred sometime after the warnings of senior colleagues to end the fascination with unsolicited border crossings. The objections which the advocates tried to head off all seemed to echo Souter's original concerns about how little of the real world was allowed into view when accepting Robbins's logical definition of methodological rigour. In section two, I show how Paul Samuelson had already brought the tendency towards explanatory unification into economic theory before it had been named as such by philosophers. The philosophy of science literature has revealed various ways in which scientific unification has proceeded, some putting causal content first, others explanatory content. But economic theory post-Samuelson always seemed to prioritize explanatory content, seeking ever more outlets for the same basic maximizing model. Mathematical analogy thus began to displace everyday economic meanings within economics. In section three, I draw out the implications of assuming for the sake of mathematical convenience that the world is more unified than it actually is. The mathematical techniques that dominate the Samuelsonian tradition enforce a purely imaginary ontological commonality onto all manner of social phenomena of very different essence. Uskali Mäki (2001a: 493) labels the ensuing explanations objects of derivational unification to distinguish them from cases where the causal content is shown to be the same across varied contexts.

ECONOMICS IMPERIALISM IN THE ACADEMIC IMAGINATION

In their important book on economics imperialism, Ben Fine and Dimitris Milonakis (2009: 11) note how often its advocates have claimed that their decisive victory was very soon to be realized. From the time of the Robbins

definition onwards, in response to his suggestion that his new analytical definition placed no substantive constraints on the application of market models'
equilibrium solutions, it has been said that the takeover of the social sciences
by economics was within touching distance. Perhaps such claims were noticeably premature in the 1930s, because the methodological argument that mathematics was a language like any other was not yet generally accepted even
within economics itself. This is what Philip Mirowski (2012b: 165) calls the
linguistic turn in the use of mathematics in economic theory, but it postdates
Robbins, being associated most obviously with Samuelson's (1983 [1947]: 12,
1983a: vi) claims that mathematization was merely another way of saying
what economists had always been trying to say about equilibrium but more
precisely (Weintraub & Gayer 2000: 441). The first generation of economists
whose textbooks encouraged them to think in this way attended graduate
school only in the early 1950s. This maybe marks the start date, then, for
economists to genuinely start appropriating the subject matter of the other
social sciences.

A quarter of a century later, though, the Nobel laureate Ronald Coase (1978:
205) was warning against expanding the boundaries of economics through
hostile activities aimed at its neighbours. There were conscious limits, he said,
to treating every aspect of human interaction as if it could be understood as
an element of market exchange. Coase (1978: 202) cut to the chase straightaway, quoting Edward Gibbon's *History of the Decline and Fall of the Roman
Empire* to emphasize the Emperor Augustus's swift acceptance of the futility
of offensive battles to secure territorial acquisitions. He argued that economists' expansionist pursuits would prove more sustainable if they sought
permission for their boundary hopping and focused on deepening combined
empirical knowledge of relevant subject matter. Souter's 1933 reflections are
not referenced in Coase's 1978 article, yet the two clearly harboured parallel suspicions of border transgressions that departed from an agreed coming
together for mutual benefit.

The early history of claims about economics imperialism therefore passes
really rather quickly from celebratory pronouncements that success was to
be expected any moment now to a rather glum assessment that a wrong turn
had been taken right from the start. In retrospect, though, a number of landmark publications from this era have been presented as evidence of economists' successful exportation of mathematical market models to other areas

within the social sciences. Amongst these well-known works are Anthony Downs's (1957) *An Economic Theory of Democracy*, Gary Becker's (1957) *The Economics of Discrimination*, Duncan Black's (1958) *The Theory of Committees and Elections*, James Buchanan and Gordon Tullock's (1962) *The Calculus of Consent*, Mancur Olson's (1965) *The Logic of Collective Action*, and Richard Posner's (1972) *Economic Analysis of Law*. Each in its own way recast aspects of legal, social, cultural or political life as the direct analogue of an abstract choice problem set in an equally abstract market environment. Each also left a deep imprint on the existing disciplinary-specific specialist literature. Social segregation, party political competition, the essence of legal statutes, the workings of parliaments, social stratification, constitutional design, group identify and all manner of institutions in which everyday life is conducted were all forced onto the same mapping of commodity space. Initially, this was a concept that allowed economists to explain from first principles the dynamics of demand and supply for any good that was produced for sale. Subsequently the mathematical derivations became increasingly unnecessary, because the imprint of demand and supply curves was left behind in the insistence that social life should be thought to observe a strict market logic (see Introduction).

A *post hoc* image of economics imperialism in full-on colonization mode is evident throughout these books. But what is not visible is any direct textual acknowledgement that this is what the authors were seeking to accomplish. Their intentions are all voiced in terms of what extra might be learnt from applying a market model to traditionally non-market-based aspects of everyday life, not via the economics imperialists' later dream of displacing existing forms of scholarship. Presumably it is to be expected that the earliest proto-economics imperialists did not cross this line. Their books were either already published or at an advanced stage of drafting before Boulding lamented the tone-deaf nature of his colleagues' attempts to redraw the boundaries of social science disciplines by calling the practice "economics imperialism". Souter's prior designation was available to them, but they can be forgiven for being unaware of it. His *Prolegomena of Relativity Economics* made only a minor dent on the professional consciousness even in its own day, and a mere generation later its claims were almost completely forgotten. Even if Downs, Becker, Black, Buchanan, Tullock, Olson and Posner had been conversant with the finer details of Souter's position, they could not in any case have been expected to adopt his meaning of "economics imperialism" as a clarion call for their activities. What they were doing amounted to a realization of his worst fears.

The earliest proto-economics imperialists therefore lacked the language of economics imperialism to describe their objectives, certainly one that carried any sort of positive connotation. George Stigler's 1984 article in the *Scandinavian Journal of Economics* was the first to offer an openly celebratory account. In 1982, Stigler became the twentieth person to be awarded a Nobel Memorial Prize in Economic Sciences. Like all newly crowned laureates, he used this elevated status to issue a series of personal statements about the future prospects for the discipline. The Nobel Committee (1982) explained that the award was for Stigler's "seminal studies of industrial structures, functioning of markets and causes and effects of public regulation". He saw market logic everywhere, not just in the operation of market institutions, and his analysis of industrial structures (Stigler 1976) and public regulation (Stigler 1971) proceeded from the assumption that they should always be understood as if they bore the characteristics of a pristine market model. This is the message he took on the road during the extended speaking tour that followed the award of his Nobel Prize.

For the lecture that provided the text for his 1984 article, though, Stigler was actually back home at the University of Chicago. When he referred to the "areas in which the economist-missionaries have ventured, often against apprehensive and hostile natives", some of the most notable of their kind were sitting in the audience in front of him (Stigler 1984: 304). The script of the lecture reads as an exercise in homage. He was clearly paying tribute to the ideas that had been in circulation throughout his storied career at Chicago, the influence of which were instrumental to him winning the Nobel prize. They also come across as a reminder of his own farsightedness in committing early to the idea that a market model is all that is needed to explain any social phenomenon. There is no hint of embarrassment, circumspection or even doubt when he asserts that "economics is an imperial science: it has been aggressive in addressing central problems in a considerable number of neighboring social disciplines, and without any invitations" (Stigler 1984: 311). Warming to his theme, he ended with a rhetorical flourish that is still unsurpassed in the tub-thumping of later economics imperialists: "[Hermann] Heinrich Gossen, a high priest of the theory of utility-maximizing behavior, compared the scope of that theory to Copernicus' theory of the movements of the heavenly bodies.

Heavenly bodies are better behaved than human bodies, but it is conceivable that his fantasy will be approached through the spread of the economists' theory of behavior to the entire domain of the social sciences" (Stigler 1984: 312–13).

Once Stigler's 1984 article had created the context for celebratory accounts of economics imperialism, others were quick to seize the initiative. Jack Hirshleifer was the first just one year later, in an over-the-top rhetorical display published by the discipline-leading *American Economic Review*. Hirshleifer's research was less well known than Stigler's, but he too began his teaching career at Chicago and, true to those roots, he exhibited a lifelong interest in extensions of price theory (Hirshleifer 1976). Economists, he lamented, "have been over-respectful" of the "confusing clamor of competing categories" that act as the other social sciences' counterpart of what, in economics, has always been a single "integrating theoretical structure" (Hirshleifer 1985: 61, 62). Hirshleifer (1985: 53, emphases in original) wrote of the two narrowing conceptions that appear repeatedly in economists' work and serve as factors of integration for a price-theoretical view of the world: "(1) of *man* as rational, self-interested decisionmaker, and (2) of *social interaction* as typified by market exchange". Whereas Stigler focused on simply reporting the fact of colonizing activity, Hirshleifer went a step further in justifying such transgressions on the basis of the innate superiority of economics. Economists, he argued, must take up the responsibility that comes with the "imperialistic destinies" of their discipline's greater explanatory successes (Hirshleifer 1985: 66). "It is ultimately impossible to carve off a distinct territory for economics, bordering upon but separated from other social disciplines ... *There is only one social science* ... [E]conomics really does constitute the universal grammar of social science" (Hirshleifer 1985: 53, emphases in original).

Next out of the blocks was Gary Becker. Not for nothing has it been suggested that he had "earned Commander-in-Chief ranking in the ... Economics Expeditionary Forces" (Demsetz 1997: 1). As Walter Block (1993) wrote following the award of Becker's Nobel Prize in 1992: "just as the fictional victims in Arthur Conan Doyle's novels trembled when Professor Moriarty was about town, almost no scholar is safe in the fields of history, law, sociology, psychology, criminology, political science, or philosophy when Gary Becker's word processor is turned on". It was these activities that occupied a large part of Stigler's and Hirshleifer's thoughts when praising the ground-breaking studies of economics imperialists. "'[E]conomic imperialism'", Becker replied when quizzed directly on the issue in a 1988 interview, "is probably a good description of what I do" (Swedberg 1990: 39). In the obligatory round of mission statements he issued after the Nobel Prize Committee had announced his award, he explained directly in the language of economics imperialism his "belief that economic analysis can be applied to many problems in social life, not just those conventionally called 'economic'. The theme of my Nobel lecture, based on my life's work, is that the horizons of economics need to be expanded ... In that sense, it's true: I am an economic imperialist" (Becker 2010). Like Souter before him, Becker spoke of "economic" and not "economics imperialism", but unlike Souter he did so to express his faith in the "combined assumptions of maximizing behavior, market equilibrium, and stable preferences, used relentlessly and unflinchingly" (Becker 1976: 5). Becker assumed that each of these characteristics comes from nature, and therefore there were no limits to where such models might act as suitable explanations for social phenomena. In a 1971 book on price-theoretical applications of market models, he told his readers that this was the "one kind of economic theory" he found permissible, and its insights could then also be imposed on a vast array of traditionally non-economic subject matter (Becker 2007 [1971]: xviii). It is presumably something more than coincidence that two of the three economists thanked for helping him to clarify this way of thinking were Stigler and Hirshleifer.

By no means everybody joined in the set-piece self-congratulations that economics imperialists now seemed so eager to indulge. Don Ross (2012: 704), for instance, has reflected thoughtfully on the "no small number of economists [who] suffer from an analogue to post-colonial guilt over their discipline's perceived arrogance as self-nominated 'queen of the social sciences'". Yet for the genuine believers, the years between Stigler and Becker receiving their Nobel prizes in 1982 and 1992 marked the moment when the genie was well and truly released from the bottle. Very quickly during this period, economics imperialism went from something that economists would not name directly to describe their boundary-hopping activities to something whose successes could scarcely be overstated. Within another ten years, Edward Lazear (2000: 140) was calling those who were still to be persuaded the new "barbarians at the gate". Lazear's account in the *Quarterly Journal of Economics* contains the usual decontextualized platitudes to the extra "rigor" that can only be delivered by the economic approach. The other social sciences can never hope to match the logical precision of a mathematical market model, he concludes, because they are obsessed with "the process rather than the equilibrium", a priority that "prevents the analyst from seeing what is essential" (Lazear 2000: 102, 128, 100).

We are today two decades further on, and from this vantage point Lazear's intemperate dismissal of the other social sciences' right to an independent existence now seems to be the high-water mark of the unquestioningly celebratory approach to economics imperialism. The literature has taken a philosophical turn since the millennium, with its centre of gravity moving from the small number of economists who revelled in the success of the imperial venture, to the admittedly equally small number of philosophers who have sought to ground such claims in an abstract account of the explanatory practices they imply. Read retrospectively, Lazear's article contains two seemingly throwaway comments that inadvertently pre-empt philosophers' concerns to learn more about economics imperialism than economists themselves have let on. He consistently targets forms of social scientific explanation that have not yet given way to the application of market models, complaining that "constraints are accepted as part of the problem's description", instead of being assumed away so that the axiomatic structure of the explanatory model can act unhindered by attention to the realities of everyday experience (Lazear 2000: 129). "The goal of economic theory", Lazear (2000: 142) concludes by way of contrast, "is to unify thought and to provide a language that can be used to understand a variety of social phenomena".

Yet this most shameless celebration of the innate superiority of economics over every other social science was being written at a time when, in a distinctly parallel literature, philosophers were already chipping away at the self-confident edifice that the economics imperialists had created to sustain their ambitions. Lazear's account places on a pedestal a form of scientific unification in which every explanation must operate on the same broad principle, whatever the subject matter. This requires the imposition of a single template where the content of each causal chain is clearly of secondary status to mathematical expedience. However, Lars Udehn (1992: 245) had already argued sometime previously that a good deal of the ensuing explanation was entirely *ad hoc*, leading at most to "a Pyrrhic victory, won at the price of an almost complete loss of substance". Uskali Mäki (2001a: 503–04) was asking economists, as a result, to reflect on whether their preferred form of explanatory unification was all it was made out to be.

EXPLANATORY UNIFICATION AND SCIENCE HISTORY

In 1949, William Kneale became the first philosopher to place at the heart of their understanding of explanation the goal of reducing the number of phenomena that each require their own independent means of analysis. He wrote an embryonic unificationist's manifesto when suggesting that explanation works best when it promotes "a simplification of what we have to accept because it reduces the number of untransparent necessitations we need to assume" (Kneale 1949: 91). In economics, Paul Samuelson (1983 [1947]: 3) was already pointing to something similar in his 1947 *Foundations of Economic Analysis* (see Chapter 6). "Only after laborious work" on many different economic issues, he confessed, "did the realization dawn upon me that essentially the same inequalities and theorems appeared again and again". The image of explanatory unification is clearly present in this comment, but the language of explanatory unification had not yet made its way from the philosophy of science to influence how economists spoke about their methodological aims. Indeed, it remained secreted in some of the furthest reaches of the philosophical literature before the general idea was rehabilitated and the process was named as such a quarter of a century after Kneale's initial intervention.

As will become apparent in later chapters (see also the Timeline), much of great significance to the prehistory of economics imperialism happened between these dates, but there was no spill-over of vocabulary from philosophy to help the economists clarify what they were doing. It would almost certainly be too much to have expected anyone in economics to have used this language prior to its second coming, this time in explicit form, in the mid-1970s. By then, the historical events I recount in Chapters 4–7 were already complete, but noticeably celebratory accounts of economics imperialism only began a whole decade later. The language of explanatory unification was in increasingly widespread use in the philosophy of science by the mid-1980s, but once again it failed to find a foothold in articles seeking to justify economists' imposition of their mathematical market models onto subject matter that had traditionally been seen to exist beyond the market domain. The case for economics imperialism rests on its most vocal advocates' contention that economics "is a genuine science" (Lazear 2000: 99). Yet at no stage did those same people think that their position would be strengthened by invoking appraisals of what is entailed in explanation from the philosophy of science. Scholars of science history, by contrast, have been much more forthcoming in seizing upon the language of explanatory unification.

We are now in a second phase of this development, where revisionist histories of science are warning that the initial embrace of the idea of explanatory unification went too far too soon. Whereas economics imperialists still invoke only a general image of explanatory unification but without talking about it directly, scholars of science history have presented the literature with accounts of the many particular ways in which such unification might materialize in practice. The proponents' case for economics imperialism has failed to move forward in any obviously discernible way from Stigler's initial exhilaration at economists' uninvited boundary-hopping activities. They continue to advocate subsumption of all legally, socially, culturally and politically relevant content to the mathematical formalisms of their market models so as to turn every issue into an abstract choice problem. Meanwhile, philosophers of science have acted upon the promptings of their science history colleagues to draw up an abstract typology of different kinds of explanatory unification. They have identified: (1) unification trends that follow (a) careful empirical study, (b) the adoption of similar mathematical formalisms, or (c) merely personal commitment to the idea of a unified world; (2) unification outcomes that are consistent with (a) known facts drawn from the world of observation or (b) stylized facts that reflect aprioristic theoretical commitments; (3) unification that results from (a) the epistemological priority of bringing clarity to everyday experiences or (b) the epistemic appeal of deductive parsimony; (4) unification that takes the form of (a) connecting a single cause to all of its demonstrated effects or (b) merely identifying one thing with another; (5) unification based on (a) a reductive logic or (b) a synthetic logic; and (6) unification whose character is variously (a) metaphysical, (b) psychological, (c) methodological or (d) logical. The discussion of economics imperialism by economists begins to look exceptionally thin when viewed through the rich variety of points of difference to be found within the philosophical literature.

However, this is not to suggest that all has always been well within the philosophical debates. Today's second phase, that of revisionist historical studies of the nature of scientific advances, serves as an antidote to an overly zealous adoption of the revived idea of explanatory unification. Once it became customary to think in such terms in the mid-1970s, there followed a period in which philosophers of science saw unification pretty much everywhere they looked (Saatsi 2017: 183). Unification quickly became established as a generically good thing, a standard to aim for and a measure of success once achieved (Hodgson 2001: 4). It appealed to the scientific ideal of being able to explain a lot with a little, as the further application of the same explanatory principles left less in the world still in search of its own unique means of analysis (Glymour 1980: 37–8). Any act of unification was treated at this time as an unequivocal advance, hinting very strongly that in future a smaller number of model types could be relied upon to bring ever more real-world phenomena into their embrace.

This appears, though, to rely on the world which unified theories are seeking to explain itself being unified, so that the same explanation works equally well for multiple real-world phenomena. Philip Kitcher (1989: 495) notes that an assumption must be in play that we are in the presence of an explanatory model "whose primitive predicates are genuinely projectable", thus enabling them to "pick out genuine natural kinds". This is what Uskali Mäki (2001a: 498–502, 2009a: 364–5) calls ontological unification. It is not just the theorists' self-made model worlds that behave in the same way and respond identically to the same causal impulses. It is the real phenomena that the model world represents which must do likewise. In the case of economics imperialism, it means that the essence of constitutional design, democratic systems and policy settings, the reason we are more willing to trust some people rather than others to run our government, the decisions of who we commit to spending the rest of our lives with, how many children we seek to have, the route we plot from job to job in pursuit of satisfying both our career aspirations and our work–life balance, our responses to legal, religious, cultural and health-related constraints on our actions, all of these must genuinely be the same kind of choice as to why we buy one particular pair of shoes and not another. Tastes and technology are all that matter. Following the lead of Stigler and Becker (1977: 76), economics imperialists routinely claim – contrary to the evidence, it should be noted – that these can be reduced to the same set of givens (see also Lucas 1981: 11–12). As a consequence, they also argue that the same market models explain everything of social interest equally well.

Yet as David Lewis (1973: 558) has asked in one of the early philosophical challenges to seeing unification wherever explanation is attempted, what can we say about explanatory unification if the world itself is not unified? Are we left looking merely at a thought experiment that entails causal misdirection? At the very least, such situations involve a different kind of projection: one where the world within the model is superimposed onto the world beyond no matter how distant the resemblance it displays to real phenomena. If multiple model worlds are made to look more alike than their real-life counterparts simply so they can be fitted to the same system of equations, then this clearly implies a different type of explanatory unification to before (Backhouse 1998a: 1852). Mäki (2001a: 493–8, 2009a: 363–4) this time calls it derivational uni-fication or, more pointedly, asks us to treat it as *mere* derivational unification to signal that its epistemic successes are nowhere near as impressive as those for ontological unification. In the case of economics imperialism, it limits the usefulness of mathematical market models to saying that they can explain all choice problems with the same degree of accuracy only if the background content of the decision-making context is understood in identical fashion by everyone in all circumstances. This is a very big "if". The model can still be relied upon to produce its answer, but it cannot be known whether the resulting solution applies to the way in which any actual people live their lives (Davis 2012: 208). The early philosophical excitement surrounding explana-tory unification was punctured when scholars of science history began to raise such doubts by showing that there were many more instances of derivational unification than ontological unification.

Isaac Newton's success in the 1680s in reducing the principles of mechan-ics to a series of fundamental equations proved to be an important historical forerunner in this regard (Scheibe 2022: 164). Newton's theoretical innova-tions enabled Kepler's laws for celestial bodies and Galileo's laws for earthly bodies to be brought under the influence of a single mode of thinking for the first time (Grosholz 2016: 63). If planetary motion and terrestrial motion could henceforth be explained in the same way, why not every other physi-cal phenomenon that could have the properties of motion overlaid onto it? For instance, eighteenth-century Newtonians had observed the outcomes of numerous experiments that made chemicals react. They sought explanatory content for these observations by proceeding as if the component elements of the ensuing compound had been put in motion by forces of attraction acting between individual particles (Kim 2003: 391–438). A variant of Newton's force laws could then be used to capture the resulting chemical reactions as if they were directly analogous to the gravitational effects that Newton's laws had explained. In derivational terms, of course, the use of the same system of equations meant that they *were* directly analogous. Likewise, eighteenth-century Newtonians had observed the outcomes of numerous experiments to project light through different-shaped surfaces. They sought explanatory content for these observations by proceeding as if the particles of light had been put in motion relative to the matter around them (Darrigol 2012: 121–35). A variant of Newton's force laws could then be used to capture the resulting optical dynamics as if they were also directly analogous to the gravitational effects that Newton's laws had explained. Once again, in derivational terms if nothing else, they were.

In the *Philosophiæ Naturalis Principia Mathematica*, Newton (1848 [1687]: lxviii) presented many explanations of physical phenomena, but always only through a single style of argument, what he called "the same kind of reasoning from mechanical principles". The unification of Kepler's and Galileo's laws therefore proceeded by first assuming that the relevant physical phenomena were in essence the same, before then asserting that an explanation which works in one instance must also work in the other instances too. An ontological shortcut was thus enacted for epistemic purposes. As Kitcher (1989: 495) has argued, this had the effect of reducing causal notions to functional subsidiaries of explanatory notions. The explanatory template was placed over various aspects of reality whose actual causes, rather ironically, remained unexplained. In most cases, he cautions, this has proved to be the furthest outreach of scientific unification: Newtonian ontological shortcuts continue to rule the roost.

Kitcher (1981: 516, 1989: 432) has developed the idea of a general argument pattern to show what is going on in these cases. The first step in its construction is to specify the relevant schematic sentences that provide insight into the problem to be solved, involving the replacement of the original non-logical expressions by dummy letters. Samuelson's (1952: 58, 1972: 12, 1977: 47) defence of bringing abstract mathematical notation to economic theory is precisely of this nature (see Chapter 6). It does not change the essence of the problem specification, he says, but brings additional clarity to it through substituting in wherever possible logical expressions for non-logical ones. This is a prime example of the linguistic turn in mathematical explanation (Mirowski 2012b: 165). The second step is to construct the filling instructions through which to persuade the audience that it is still a real phenomenon under consideration and not merely the logical expressions that are required to stand in for it when the real phenomenon is forced into a corresponding mathematical formalism. A general argument pattern will eventually be founded on what mathematicians call a set of sets of filling instructions, built up sequentially so that no schematic sentence is denied its own particular set of filling instructions (Kitcher 1981: 516). Collecting all of the relevant schematic sentences together in one logical sequence produces a schematic argument. The final step is to list the classification for the relevant schematic argument, which helps to distinguish those inferences it will permit from those it will not. The inferential characteristics of a schematic argument might become increasingly straightforward to identify through its repeated use across multiple domains, but this does not mean that those explanations would have been granted authoritative status before the general argument pattern became embedded in professional practice.

The task ahead is to decide how best to characterize economics imperialism. Does it deliver genuinely new knowledge in rigorous form to real phenomena whose essence, at most, has only previously been partially understood by its subject specialists? Or has it succeeded only in showing us how those phenomena would be expected to behave were they aspects of a general argument pattern that belongs only to a recognisably market world? The small group of explicit cheerleaders for economics imperialism is adamant that the former is true. Yet this is despite considerable evidence that the filling instructions in use for fleshing out the general argument pattern of mathematical market models do not conform to basic observations. The fears of those on the receiving end of these intellectual land grabs is that empirical rejoinders might never be enough to stem the tide of boundary-hopping activities. The broad deference to mathematical formalism could always be sufficient for the economics imperialists to win out, especially when their case is made in such confident terms. However, further discussion of the general argument pattern of mathematical market models might go at least some way towards settling this issue. Do market models permit inferences only in relation to their own internal characteristics rather than to the world beyond? What role has the mathematization of market models had, not only in replacing non-logical with logical expressions within these models, but also in replacing real situations of actual experience with hypothetical lives that cannot be lived?

MATHEMATICAL FORMALISM AND DERIVATIONAL UNIFICATION

Edward Lazear's *Quarterly Journal of Economics* article from 2000 has entered the folklore of both those seeking to further the cause of economics imperialism and those seeking to resist it. It acts simultaneously as the greatest hurrah of the triumphalist tendency and the source of the greatest indignity for anyone who has been told that their subject specific expertise is now worthless. The philosopher Margaret Morrison also published her book, *Unifying Scientific Theories*, in 2000. At no point does it mention economics imperialism, economists' market models or even economics more generally. Morrison's thoughts consequently feature directly in neither the dreams of the colonizers nor the nightmares of the colonized. Yet her book can provide foundations for a very different understanding of the limits of economics imperialism compared to anything that can be said when taking the Stigler–Hirshleifer–Becker–Lazear position at face value. *Unifying Scientific Theories* is full of insightful histories of how a common mode of explanation might come to exert great pressure, forcing different theories to become more and more alike. There are consequently important clues contained within its pages for how best to adjudicate between the various claims in the economics imperialism debate.

Morrison's (2000) detailed historical case studies focus on separate episodes of explanatory unification in the physical and biological sciences. She uses these to side with other philosophers of science such as Bas van Fraassen (1980: 87, 109) and Nancy Cartwright (1983: 12, 53) in seeking to delink unification, explanation and truth. Whatever the source of metaphysical claims concerning the inherent unification of nature, Morrison (2000: 235) says, science currently demurs. Contemporary scientific practice produces no grounds beyond aprioristic belief to think that there are a small number of all-embracing models – some already discovered, others yet to be – that will explain all we want to know about everything. Todd Jones (2001: 1097) describes faith in the existence of such models as "the common picture of the world" bequeathed by Enlightenment scholarship, and it certainly fires the passions of the most committed economics imperialists.

Far from burnishing scientific models with extra causal precision, Morrison (2000: 253) argues that the introduction of ever more detailed mathematical specification provides only a basis for talking in highly abstract terms about the state of the world as it exists within the models themselves. In Kitcher's (1981: 519, 1989: 434) terms, it allows a series of claims to be derived from a general argument pattern, but without the requirement to outline the actual causes in play. Being able to describe the course of planetary motion in mathematical terms is not the same as having isolated the causes of that motion for any particular body. Being able to describe the course of evolutionary adaptation in mathematical terms is also not the same as having isolated the causes of that adaptation for any particular organism. There is explanation going on in both of these cases, but it is not causal explanation as it is most readily understood. What, then, might be said about the type of explanation that follows from using a mathematical market model to explain non-market phenomena?

Morrison's discussion of the development of James Clerk Maxwell's laws of electrodynamics provides important awareness of what is at stake in this regard. Maxwell is held in such high esteem that his fellow physicists typically talk about him in the same breath as Newton and Einstein (Arianrhod 2005: 7). He was arguably without peer as a unifier, showing in one of the most far-reaching scientific accomplishments ever that electricity, magnetism and light could be understood as different manifestations of one and the same thing (Dharma-wardana 2013: 119). The Nobel Prize-winning physicist, Richard Feynmann, has described this as the single most important event of the age, for both the development of scientific knowledge and the way we have subsequently come to live our lives (Flood 2014: 3). However, Morrison (2000: 4) asks us to consider the *means* through which the unification was demonstrated, rather than just the demonstration itself.

This is perhaps one of those historical stories where it is best to start at the end. Physicists noted very soon after the publication of Maxwell's two-volume *Treatise on Electricity and Magnetism* in 1873 that something seemingly very important was missing from it (Maxwell 1873). Within four years of his death, Oliver Heaviside (1883: 153) expressed concern that the theory which might help us to see how the explanatory unification mirrored a real physical unification remained entirely implicit: he drew attention to "the difference between the patent and the latent, in Maxwell". Maxwell's famous equations had provided reason to proceed by assuming that electricity, magnetism and light were common phenomena, but no reason to know why in physical terms. Unification, explanation and truth once again appear to be a troubled triumvirate. There can be no doubting the scale of Maxwell's achievements, as he did nothing less than force scientists to revise their view of physical reality (Auxier & Herstein 2017: 74). He changed not only what scientists believed they could see in nature, but also the conceptual lens that provided them with the means of looking for it (Henderson 2018: 144). But this had the effect of relegating actual observation of real phenomena to a subsidiary role in the explanatory schema.

It was not always like this, though, because Maxwell's final position emerged from the second of his two transformative papers on electrodynamics (Longair 2016: 114). The first was a four-part paper published in 1861 and 1862, "On Physical Lines of Force". This built on an earlier but more rudimentary two-part paper from 1855 and 1856, "On Faraday's Lines of Force" (see Maxwell 1856, 1861–62). The lineage stretching back to Michael Faraday is important, because the early Maxwell took from him the need to provide reasons for the production and propagation of electromagnetic waves (Morrison 1994: 369). Faraday had founded his theory of electromagnetic induction on experimental facts, using observations of the effects that followed when transient currents were created under laboratory conditions (Forbes & Mahon 2014: 53–68). Maxwell initially worked within the context established by these experimental facts to present Faraday's observations in generalized mathematical form (see Chapter 3 on experimentation). This involved constructing a simple mechanical model of the electromagnetic field so that the physical phenomena to be explained continued to have a real presence within the explanation. Daniel Siegel (1991: 56) has written of the "ontological intent" that Maxwell had vested in his initial pen-and-paper model of molecular vortices. The mathematics was something to be added at the end as a clarity-enhancing adornment to the model, not the source of the explanation itself. However, this sense of priority was flipped in Maxwell's 1865 paper, "A Dynamical Theory of the Electromagnetic Field" (see Maxwell 1865).

Morrison (2018: 32) describes the general argument pattern of the later paper as an irredeemably top-down affair, in contrast to the bottom-up approach exhibited just four years earlier. Gone was the use of evidence derived experimentally to inform the construction of the model. Gone too was the idea of only then redescribing the main features of the model mathematically (Miller 1984: 120–21). In its place Maxwell substituted a mathematical model whose dynamic features followed its own inherent properties, rather than attempting to replicate observations relating to the physical phenomena the model purports to describe. To this he then affixed his laws of electrodynamics, as if the ability to show that they could be captured through a system of equations was evidence enough that the real phenomena represented in the model had been explained. The intuition might have been sound, but it is easier to say that now than it would have been in the 20 years that elapsed before Heinrich Hertz provided experimental proof of the existence of electromagnetic waves in the late 1880s. At the time of Maxwell's grand act of unification, at most he had proved the reliability of an existence theorem: the right to assume that electromagnetic waves might conceivably take the form described by the system of equations. But this is explanation without demonstration, explanation shorn of conventional truth claims, the assertion of realities – to quote Tina Young Choi (2016: 137) – that were "at once imaginable yet incalculable" (see Chapter 3 on models as thinkable entities). Hertz (1896: 318) got right to the heart of the matter when writing: "It is impossible to study this wonderful theory without feeling as if the mathematical equations had an independent life and an intelligence of their own, as if they were wiser than ourselves, indeed wiser than their discoverer, as if they gave forth more than he had put into them". Maxwell himself almost confessed as much. He suggested that often the only way to bring physical reality to life is to imagine its presence mathematically, referring to this process as "the still more hidden and dimmer region where Thought weds Facts" (Maxwell 1890 [1870]: 216).

The catalyst for the change was the introduction of Lagrangian formalism into the second of Maxwell's two presentations of the laws of electrodynamics. It enabled the system of equations to overpower in the explanation the physical phenomena they supposedly only represent. The 1861 paper had been based on twenty differential equations, each designed to add general explanatory potential to one of Faraday's experimental facts. The 1865 paper used Lagrangian formalism to take the argument to a still greater level of generality, but this time by removing all commitment to observation of real physical entities. Maxwell's famous equations *became* his theory.

Joseph-Louis Lagrange's (2009 [1788]) two-volume *Mécanique Analytique* of 1788 (in English, *Analytical Mechanics*) was one of the great integrationist achievements of the late eighteenth century. Lagrange traversed the boundary between what would later become known as pure and applied mathematics, transforming Newtonian mechanics from a branch of algebra to a branch of analysis (Rao 2011: 49). Mechanical principles were henceforth not something to be calculated via observation, so much as derived via the application of a new technique of variational calculus (Fraser 2003: 361–3). Physical explanation was thus released from an account of Newtonian forces, in which each particle had to have ascribed to it an independent acting force that might propel an entity into motion. Lagrange's revolutionary step was to treat the elementary relations of dynamic physical systems as if they were indistinguishable from the purely analytical mathematical objects used in their description (Grabiner 1996: 138–40). Within Newton's *Principia*, physical quantities such as velocity, momentum and force had to be calculated directly within the equations of motion. Not so within Lagrange's *Mécanique Analytique*, where the quantitative determination of the field could proceed without first having ascertained through calculation the motion or even the very nature of the system under study (Morrison 2018: 32). Maxwell subsequently followed exactly the same approach and made exactly the same excisions. He used Lagrangian formalism to create his system of equations while bypassing the need to know anything about the mechanical structure of the electromagnetic field or any of its physical components (Gunn 2013: 78). Newton's synthesis of Kepler and Galileo already bore many of the hallmarks of derivational unification; Lagrange's subsequent reworking of Newton could *only* be about this.

These insights assist my appraisal of economics imperialism, because Lagrangian techniques were brought with ever greater flourishes to economics post-Second World War (see Chapters 6 and 7). They introduced far greater mathematical precision to market models than ever before, but as the example of Maxwell's two papers on electrodynamics shows, they also reconfigured the balance between what could be said about the world beyond the model and what could only be said about the world within the model. It moved decisively towards the latter.

Samuelson was the first to bring systematic Lagrangian instincts into the discussion of market models. In the mid-1990s he was still able to identify the exact moment six decades earlier when this shift first took shape in his mind.

He recalled a conversation with Gilbert Bliss, the mathematician teaching the calculus class he attended at Chicago in 1933–34. "When I went to him to show how I solved a famous problem in duopoly … he said: 'Oh, you could have used Lagrange multipliers.' And presto, all mystery evaporated. I knew I wanted more of that good stuff. And like Oliver Twist, always it was a case of MORE" (Samuelson, cited in Backhouse 2017: 67). Samuelson's (1983 [1947]: 454) subsequent concern was to show how the same way of thinking could be used over and over again in perfunctory repetition so as to distil the essence of all theoretical questions in economics. "Lagrangian multipliers", he wrote in *Foundations*, "have the character of dual or shadow prices" (Samuelson 1983 [1947]: 454). Once a structure of prices is imposed where once it did not exist, everything of social interest can subsequently be turned into a market problem that might be solved in an identical manner. The schematic sentences of Kitcher's general argument pattern were thus to be represented in Lagrangian terms, and the filling instructions were to be given economic sounding names consistent with equilibrium pricing problems, so that the classification of relevant cases aligned with generally accepted principles of economic theory. As a result, real economic phenomena were increasingly displaced to a realm beyond the explanatory structure. Samuelson (1983 [1947]: 231, 60) admits to the "artifice of Langrang[i]an undetermined multipliers", acknowledging that when using them the "economic interpretation will be brought out later", only once the system of equations has been solved. The world beyond the model seems to have gone increasingly missing within the model.

The trailblazers of general equilibrium theory in the 1950s – Kenneth Arrow, Gérard Debreu, Lionel McKenzie, Leonid Hurwicz and Hirofumi Uzawa – accelerated the trend that Samuelson had started. Whereas he sought to expand the range of feasible mathematical applications within economic theory, their aim was instead to take market models to ever more rarefied heights of mathematical sophistication. Mathematicians had spent the century and a half following the publication of *Mécanique Analytique* attempting to push Lagrange's revolution further. The early general equilibrium theorists had more mathematical training and, indeed, greater mathematical skill than Samuelson. Samuelson changed the face of economic theory with his adoption of what, in mathematical terms, are rather rudimentary Lagrangian multipliers. Arrow, Debreu, McKenzie, Hurwicz and Uzawa changed it again, placing even more distance between the world within the model and any real-world dynamics it could be said to represent. They used bespoke techniques within the most up-to-date Lagrangian tradition to solve mathematical accounts of economic problems that were so abstract they lacked all but the flimsiest connection to everyday experience (see Arrow & Debreu 1954; McKenzie 1954; Arrow & Hurwicz 1958; Arrow, Block & Hurwicz 1959; McKenzie 1959; Debreu 1959; Uzawa 1960).

All rigorously specified Arrow–Debreu economies are now constructed as thought experiments on the basis of a Lagrange multiplier method that guarantees equilibrium results within the associated market models (Aliprantis 1996: 34). But to ensure that their systems of equations would be mathematically solvable in an apparently meaningful manner, the general equilibrium theorists of the 1950s had to integrate into the Lagrangian framework Shizuo Kakutani's fixed-point theorem for multi-valued functions (Hurwicz 1987: 289). Kakutani's (1941) game-changing paper was published only in 1941, some years after Samuelson's epiphany following his conversation with Bliss. Debreu and McKenzie had first been alerted to the potential for using its insights to prove the existence of general equilibrium when reading a 1950 working paper entitled "Lagrange Multipliers Revisited" by the Cowles Commission's resident mathematician Morton Slater (1950) (see Düppe & Weintraub 2014: 89). Aleksandr Lyapunov had developed Lagrange's theory so that it was possible to further specify the stability conditions of dynamic mathematical systems, and Lyapunov–Lagrange stability plus Kakutani's fixed-point theorem provided the general equilibrium pioneers with all the mathematical structure their economic theory required (Dominique 2001: 49–50). Clearly, there is no concern here for real economic phenomena. The irony resurfaces that the colonization of the social sciences by economics passes through the elimination of economic content from so much of economic theory (see Introduction). We are being asked to trust a mode of thinking that seems to have deprived itself of the means of answering its own core questions to nonetheless answer everything else (Aydinonat 2015: 56).

It is interesting in its own right to piece together the chronology of the mathematical developments that have been incorporated into various market models. But this is not what is most important for current purposes. Samuelson's attempts to generate the widest possible applications for mathematical market models clearly inhabit a different historical trajectory to Arrow,

Debreu, McKenzie, Hurwicz and Uzawa's efforts to raise market models to the highest possible levels of mathematical purity. Yet in each case the underlying style of argument is the same, and this is what matters most for the chapters to come. The common style of argument is indicative of what Morrison (2000: 5, 232) calls synthetic rather than reductive unity and Mäki (2001a: 493–502, 2009a: 363–5) derivational rather than ontological unification. The former of these pairs signals only that multiple phenomena have been grouped together under the same explanatory structure, the latter that the associated real phenomena have been shown to be genuinely the same at the level of causation. An abstract system of equations can be used to line up a number of disparate entities in apparent conformity, whereby "rules springing from remote and unconnected quarters should thus leap to the same point", to quote the nineteenth-century mathematical unificationist William Whewell (1858: 88). However, the philosopher Jaakko Kuorikoski (2021: 190) suggests that at most this leads to "cases of pseudoexplanation", explanation-like reasoning based on formal understanding. This allows us to say what the world would look like were it to follow the structure of the mathematical model closely, but otherwise it leaves actual causes unaddressed.

CONCLUSION

Any moment of scientific unification is still generally considered to be the sign of a growth in human knowledge. The act of using the explanation of one thing to also explain something else is suggestive of a higher plane of learning in which it is no longer necessary to start from scratch every time a new aspect of reality imprints itself upon our consciousness and demands comprehension. Or is it? This chapter has highlighted the need to beware running too far ahead with arguments of this nature.

Philosophers of science are certainly more cautious today in celebrating the virtues of scientific unification without first asking qualifying questions about exactly what has been achieved. Their science history colleagues have shown how often the most important analytical moves are confined to the mathematical structure of the explanation, where the essence of the entity under investigation is conveniently remoulded so that it proves to be a better fit for the mathematical model. The ensuing explanation is clearly of a different order to one where the mathematics is introduced only after thorough empirical study has revealed a series of core facts about how the target phenomenon works. There are consequently two dimensions to the claim that a moment of scientific unification has been secured. First, can the audience be persuaded that they are in the presence of an explanation that is successful in its own terms? If they can, but if no more can be said than this, then at most we would seem to be looking at a situation of derivational unification. Second, how often is it possible to go further and persuade the audience that what has been explained is the phenomenon as it is experienced in everyday life? This corresponds to the more exacting standard of ontological unification.

The current revisionist stage of the science history literature makes it clear that we are much more likely to witness moments of derivational rather than ontological unification. When first making the case for bringing mathematical applications directly into economic theory, Hermann Heinrich Gossen (1854) added an important qualification. This was in his *Entwickelung der Gesetze des Menschlichen Verkehrs, und der daraus fließenden Regeln für menschliches Handeln* (translated into English as *Development of the Laws of Human Intercourse and the Consequent Rules of Human Action*). He immediately drew his readers' attention to the fact that the systems of equations he would be borrowing from the physical sciences were devised for entities whose essential features were considerably more regular than those of any social organism. But he pressed on regardless, activating a tendency to ignore the ontological constraints on scientific unification that has become such a marked feature of the practice of economics imperialism. Gossen's discussion shows that he was fully attuned to the misdirection involved in attributing to human life a predictability he knew it could not have simply because it made the mathematical applications easier to perform. Stigler was sufficiently astute as a historian of economic thought to also have been fully aware that Gossen knew this about his own work. Still, though, Stigler was happy to enlist Gossen as an intellectual inspiration for celebratory accounts of economics imperialism 130 years after the publication of the *Entwickelung der Gesetze des Menschlichen Verkehrs* in 1854. There is consequently a significant parallel between Gossen first telling his fellow economists to import equations from physics to transform their market models and Stigler first telling them to use their market models to colonize the other social sciences. Everyone seemed to know from the start that they could not allow ontological constraints on scientific unification to get in their way.

It is therefore necessary to exercise caution and not invest too much in trying to dismiss the work of economics imperialists simply on the grounds that it fails the test of ontological unification. How could it be otherwise when, from these dual moments of origination onwards, it has always been such a badly kept secret that the hypothetical agents of mathematical market models will be granted a predictability that is untouched by changes to legal, social, cultural and political norms? This is not to say that there was a complete absence of attempts in the prehistory of economics imperialism to make the ideal market agent as real as possible (see Chapters 4 and 5), only that the mathematical market models in use across the social sciences today were long ago purged of such an effort (see Chapter 6). Economists still typically present their models as if they speak directly to the real world, but they struggle to say why this is the case (see Chapter 3). The most revealing critique of economics imperialism is therefore not necessarily that its explanations appear lacking when judged by the standards of ontological unification, because there is no real reason to expect them to perform at this level. It is working out what limitations on explanation were accepted when decisions were taken to settle for derivational unification. Economics imperialists might still argue that their explanations are better than those of the subject specialists they are seeking to displace from the literature, but focusing on these limits tells us what we can expect "better" to mean in such circumstances.

Economics imperialism very consciously promotes a mathematical theory of model market conditions over the empirical study of socially situated knowledge concerning actual human interactions. The explanation – or, indeed, the pseudo-explanation – all takes place within the process through which structure is imposed aprioristically on the mathematical market model. The explanatory information therefore comes from what is most convenient for the solution to the system of equations, not from a detailed account of the world that exists beyond purely epistemic practices of explanation. There remain good reasons to continue exploring the logical properties of mathematical market models as guides to how far from idealized conditions current experiences of life are. Care has to be taken, though, not to overclaim what this can tell us beyond the idealizations on which such models are based. This lack of integration between the world within the model and real-world social dynamics is the major point of philosophical weakness for economics imperialism. Yet it can itself be a significant source of concern for those who feel under threat of colonization (see Introduction and Conclusion). The more gung-ho the advocate, the more likely they are to blithely ignore the fact that the most their style of economics imperialism has to offer is synthetic reduction, and the more likely they also are to try to turn the conversation immediately to who has the best mathematical credentials. They will often do anything they can to deflect the debate away from who has the most secure grasp of empirical detail, because this is not what matters to them.

The following chapter asks about the origins of the faith in mathematics to explain social phenomena whose essential characteristics are not easily enumerated. If the contemporary practice of economics imperialism is all about imposing mathematical market models as explanations for non-market phenomena, then its prehistory must also involve the imposition of mathematical reasoning tools onto non-mathematical domains. Economics imperialists suggest that there is something special about their mathematical market models that gives them universal applicability, but this is only ever a second-order argument. First it must be believed that there is something extra special about mathematical explanation more generally, whereby mathematics can be trusted to be the arbiter of its own truth claims. Interestingly, Gossen's injunction that mathematical explanations should be allowed to stand on their own in economic theory preceded by almost half a century mathematicians' willingness to make the same argument for their own discipline. A philosophical shift occurred around the turn of the twentieth century, under which the metamathematical claim began to be accepted that if an axiomatic theorem was constructed in logically flawless fashion then it must also be mathematically true. Mathematical objects henceforth could be treated as if they had the capacity to demonstrate their own truth, rather than only being provisionally true pending verification through calculation or experimentation. These metamathematical moves have yet to attract any attention in appraisals of economics imperialism, but they are the focus of the following chapter.

CHAPTER 2

True in the model versus true in the world: mathematical postulation, proof-making activities and economic theory

INTRODUCTION

Sitting in a refreshment room at Berlin railway station in 1891, the 29-year-old David Hilbert uttered one of the most memorable lines in the whole history of mathematics. He had been attending the annual meeting of the German Mathematical Society in Halle, and he was waiting with some fellow conference attendees for his connecting train back to Königsberg. The friends were reflecting on what they believed to have been the most important talk they had heard at the conference. Hermann Wiener had delivered a provocative lecture in which he outlined the need for more rigorous underpinnings to the theory of geometry. Very few people still took Euclid's *Elements*, written in the third century BCE, as a repository of literal truths, but the primitive elements of geometry had remained largely unchanged since that time (Henderson 2013: 101). *"Man muß jederzeit an Stelle von 'Punkte, Geraden, Ebenen' 'Tische, Stühle, Bierseidel' sagen können"*, Hilbert suddenly interjected into the conversation (Blumenthal 1935: 402–03). This starkly revealing sentence is usually translated into English as: "You can say at any time 'tables, chairs, beer mugs' instead of 'points, straight lines, planes'".

Such is the retrospective power that has been loaded onto Hilbert's comment that it sounds as though it belongs to an apocryphal story. However, the biographer in question, Otto Blumenthal, was one of Hilbert's closest collaborators. His account was seen by Hilbert before it was despatched for publication, presumably therefore with his blessing. There is still the chance he did not use these exact words and the standard English translation might add a more dramatic gloss to what was actually said. At this stage of his career, though, Hilbert certainly thought it should be possible to substitute any words drawn from a random letter generator for the well-known geometrical concepts of "points", "lines" and "planes", yet still leave intact the underlying logical structure through which he would henceforth seek to describe the relationship between points, lines and planes. I know of no contemporary economics imperialist who cites Hilbert's 1891 challenge to the need for confirmed empirical content within mathematical objects as direct inspiration for their activities. But the parallels appear to be more than merely suggestive. Hilbert's railway station epiphany suggests that it is the identification of the correct mathematical procedure that leads the proofs to the solution, not knowledge of the real phenomena to which the proofs apparently relate. What holds for mathematics, it seems, subsequently also holds for mathematical economics and, from there, for economics imperialism too. The justification for imposing mathematical market models onto non-market entities in the legal, social, cultural and political domains is of the same basic form as Hilbert's Berlin pronouncement of how best to rework the foundations of geometry.

This, after all, had been the essence of the challenge that Wiener had laid before his audience. He had insisted that geometry should be studied at one stage removed from the visual images that helped to make otherwise less than fully defined geometric propositions "obviously" true within the mind's eye (Gandon 2016: 48). He told his fellow mathematicians to go out and think about how best to enact a radical break with the reasoning tools handed down by Euclid. "Tische, Stühle, Bierseidel" was thus Hilbert's code for fleshing out Wiener's suggestion that a system of carefully specified axioms could replace the need for empirical statements that defined geometrical properties in visual terms. Primitive geometrical concepts could only be placed on a rigorous footing, he would go on to argue, within the context of a logical structure which was internally consistent and allowed for no basic contradictions caused by intransitivity (Toepell 2005: 714). Hilbert's unexpected station platform utterance has become so legendary because it says that it no longer matters exactly what counts in substantive terms as a primitive geometrical concept. All that is of interest post-"*Tische, Stühle, Bierseidel*" is that the logic describing their internal relationships is sufficiently skilfully elaborated as to satisfy the conditions for a successful axiomatic system (Koetsier 2002: 203). This is the sanctioning of what the philosopher of mathematics Stewart Shapiro (2009: 437) has described as "freestanding" systems. Knowledge of Hilbert's role as a metamathematician is very rare amongst practicing economic theorists, but they often appear to have been converted wholesale to the epistemic value of mathematical objects that follow Hilbert's prescriptions.

However, the more mature Hilbert was rather usurped by another young mathematician, the 24-year-old Kurt Gödel, before he could flesh out his axiomatic programme in its entirety. Gödel revealed the first of his two famous incompleteness theorems at the Second Conference on the Epistemology of the Exact Sciences held in Hilbert's old stomping ground of Königsberg in 1930. He was the least likely of the roundtable participants to have been the main draw for the people who heard him speak on 7 September. Yet his short statement was to leave an indelible imprint on mathematical practice. Hilbert himself was in town that day, preparing for his own invited lecture on 8 September to the Society of German Scientists and Physicians (Sieg 2012: 96). Hilbert's lecture on the Monday, "Naturerkennen und Logik", took his own foundational propositions for granted, but Gödel's contribution to Sunday's roundtable had severely undermined them. Gödel had proved that the logical starting point for Hilbert's formal axiomatic reconstitution of mathematics was itself logically insecure.

In order to pursue further the significance of this finding for my study of economics imperialism, the chapter now proceeds in three stages. In section one, I start with where the Hilbert programme might well have finished, which is with the publication of Gödel's incompleteness theorems. I ask why those theorems simultaneously represent a remarkable mathematical achievement but had only limited impact metamathematically in dethroning the Hilbert programme. In section two, I look behind the Hilbert programme to see where he fits into the history of the rigour movement which aimed to put mathematics on surer, more precise foundations. Hilbert was only one of many metamathematicians whose work signalled advances in the rigour of mathematical proof-making at the time that economists' basic market models were being reconstituted mathematically. In section three, I show how methodological pluralism on questions of rigour remains the default setting even in mathematics, from the outside by far the most unified of academic disciplines. The language of rigour has become a standard feature of what economics imperialists believe their market models bring to other social sciences. Yet this appears to pay no heed to where exactly within the history of metamathematics such claims to rigour are situated. It is not the unproblematic single entity that it needs to be for the arguments of contemporary economics imperialists to survive unchallenged.

GÖDEL'S INCOMPLETENESS THEOREMS

Hilbert's work can be seen as an end-point of two thousand years of striving for a genuinely axiomatic treatment of geometry that might be extended to encompass the whole of mathematics (Chaitin 1995: 89). Not without reason, then, has it been said that he should be raised to the "rank of an Archimedes, Newton, or Gauss" (Leonard 1992: 40). If foundational certainty is the key to developing a mathematics no longer reliant on basic intuitions drawn from common sense, then the Hilbert programme that was flourishing by the late 1920s looked as though it was successfully providing it. By locating the foundations of his subject field in an axiomatic system which abides by the formal rules of logic, Hilbert (1902: 444) was sure that he could propel mathematics to a situation in which "every definite mathematical problem must necessarily be susceptible of an exact settlement, either in the form of an actual answer to the question asked, or by the proof of the impossibility of its solution and therewith the necessary failure of all attempts". He reformulated the basic principles of geometry by saying that lines should henceforth be viewed as sets of points. The move from geometry to arithmetic was deceptively straightforward, because what was true for lines in general could also be considered true for the specific case of the real number line. The logical rules of arithmetic could therefore be made to reflect fixed entities out of which the set of real numbers is comprised (Stewart & Tall 2015: 259).

Gödel's (2003 [1970]: 10) incompleteness theorems arose from work intended to assist in this endeavour, by counteracting the "philosophical prejudices" to which the Hilbert programme was currently being subjected. However, far from demonstrating that every mathematical problem could elicit a provable solution, Gödel showed that some foundational issues in mathematics remained fundamentally unprovable as a purely mathematical matter.

The incompleteness theorems begin with a question that sounds innocuous enough: how might we know that a mathematical statement we are sure is true of all currently known examples is also provably true for all possible examples (Dawson 2008: 819). Hilbert's (1902: 478–9) metamathematical work was based on the assumption that mathematics stood alone as the only field of knowledge that could be trusted to be the arbiter of its own validity (Corry 1989: 413, 2004: 4). No external benchmarks were necessary; self-validation was the key. Hilbert was well aware that this meant it had to be the case that every mathematical statement could be proved to be true (in all instances) or false (in at least one), provided there were no holes in the surrounding formal axiomatic system. Gödel's intention was to add extra foundational rigour to such a claim.

Hilbert knew that the big prize in this regard was the axiomatization of arithmetic. Logicians talk about the negation-completeness of a theory T if either ϕ or its negation $\neg\phi$ can be derived from within T's system of proof for every sentence ϕ that occurs in the language of the theory (Smith 2013: 2). To take a very elementary example, we are likely to have few problems convincing ourselves that 3 plus 4 does not equal 12 but that 3 multiplied by 4 does. No recourse to the use of logical axioms will be necessary to determine whether we are right. We will have been conditioned by rote learning in childhood to know instinctively that 3 plus 4 feels wrong if we are looking for the answer 12, but that 3 multiplied by 4 feels right. However, the whole objective of the Hilbert programme was to release mathematics from reliance on this sort of intuitive feeling and to refound it instead on a logical rather than a psychological basis.

It is not difficult to show how this might be done for such a simple example. The number sequence of basic arithmetic allows for no jumps of more than a single integer at a time; every number is exactly one greater than that which precedes it; there is no highest allowable number that brings the sequence to an end; and there are no numbers that exist somewhere beyond the parameters of the sequence. This allows us to construct straightforward rules which are logically sound. The feeling that $3 + 4 \neq 12$ can be formulated as a logical rule for addition stating that, to add b to a, you have to move b places along on the number sequence having started at a. It is just as easy to respecify the feeling that $3 \times 4 = 12$ as a logically sound rule. To multiply $a \times b$ it is necessary to begin at zero in the number sequence and then add a on a repeated basis until you have done so *b* times. The ultimate success of the Hilbert programme rested on the possibility that, once the position of negation-completeness had been reached, every statement relating to number theory should be provable in either the positive or the negative, even if it had not yet been proved. From Hilbert's perspective Christian Goldbach's famous conjecture that every even number from 4 onwards could be reduced to the sum of two prime numbers currently belongs in the category of the provable but as yet unproven (Vaughan 2016: 479). However, provable but unproven clearly belongs to a different category of truth to proven.

By 1930, Hilbert was talking excitedly about the possibility that mathematics had already demonstrated the consistency of number theory with just one or two minor exceptions. Hermann Weyl's wish for mathematics to become the science of ∈, meaning set membership, seemed very close to fulfilment (Ferreirós 2007b: 339). Mathematics appeared to be on the cusp of a complete break with the idea that it was the science of quantities (Archibald 2008: 128). When those final few steps had been taken, Hilbert assumed, it would follow that the axiomatic system would be capable of fixing the truth-value of any mathematical statement that emerges from the use of the language of arithmetic (Zach 2007: 418). For every proposition ϕ, the relevant deductions from the axioms would be able to prove either ϕ or $\neg\phi$, thus demonstrating that the initial proposition was either true or false. Imagine the sense of shock that must have pervaded the room, then, during Gödel's 20-minute slot at the 1930 Königsberg Conference on the Epistemology of the Exact Sciences. This is where he first gently suggested that mathematicians were not currently in a position to claim the completeness of the axiomatization of arithmetic and, what was more, there was good reason to presume they never would be.

Perhaps more tellingly, we should imagine the shock of those people who understood the full implications of what Gödel had just said. Reports from those who were present suggest that most participants were unmoved in the moment, because the highly abstract nature of Gödel's pronouncement, coupled with the understated manner in which he made it, meant that the spectre of crumbling foundations passed most people by (Murawski 1999: 202–03). Certainly, word does not seem to have got back immediately to Hilbert. When Gödel first showcased his results, he told his audience very quietly and without any hint of grandstanding: *"Man kann (unter Voraussetzung der Widerspruchsfreiheit der klassischen Mathematik) sogar Beispiele für*

Sätze (und zwar solche von der Art des Goldbachschen oder Fermatschen) angeben, die zwar inhaltlich richtig, aber im formalen System der klassischen Mathematik unbeweisbar sin" (Gödel 1931: 148). In English, this translates as: "One can (assuming the consistency of classical mathematics) even give examples of propositions (and indeed, of such of the type of Goldbach or Fermat) which are really contentually true but are unprovable in the formal system of classical mathematics" (Dawson 1984: 126). On the very next day, Hilbert spoke as if nothing of any importance had happened just around the corner only 24 hours previously. If word had reached him about Gödel's address it did not cause him to alter the content of his own. He issued a rallying call to his fellow mathematicians via an explicit refusal of the physiologist Emil du Bois-Reymond's idea that scientific knowledge is necessarily limited: *ignoramus et ignorabimus* ("we do not know and shall never know") (see Rescher 2006: 101). "*Wir müssen wissen. Wir werden wissen*", countered Hilbert (1930b: 963) to the "foolishness" of such a proposition ("We need to know. We will know"). The contrast was thus very clear: complete confidence in the act of knowledge creation from Hilbert versus the quiet demonstration from Gödel of the limitations of willing the indeterminable.

It is impossible to miss just how meticulous Gödel was when giving initial notice in Königsberg of the results on which he was sitting. There was no presentational flourish or clever play on words designed to make his audience sit up and listen. Perhaps this was why the significance of what he was saying seemed to be lost on most people in attendance (Goldstein 2005: 157). Forty years later, Gödel (2003 [1970]: 10, emphases in original) elaborated on those results in the following way, still without any rhetorical arm-waving:

> [A]n arithmetical model of analysis is nothing else but an arithmetical
> $\in$-relation satisfying the comprehension axiom:
> $$(\exists n)(x)[x \in n \equiv \phi\,(x)].$$
> Now, if in the latter "$\phi\,(x)$" is replaced by "$\phi\,(x)$ is provable," such an
> $\underline{t}$-relation can easily be defined. Hence, if truth were equivalent to
> provability, we would have reached our goal. However (and this is
> the decisive point) it follows from the *correct* solution of the semantic
> paradoxes that "truth" of the propositions of a language *cannot be
> expressed* in the same language, while provability (being an arithmetical relation) *can*. Hence true $\neq$ provable.

Gödel was interrupted only once when delivering his Königsberg address. This was a clarification question from John von Neumann, who has typically been seen as Hilbert's *de facto* representative at the conference (Mathias 2014: 59). As the Gödel scholar John Dawson (1986: 197) has noted, Gödel then changed register "abruptly, as though emboldened and spurred by von Neumann's leader". Gödel's (1931: 148) upshift saw him transition from subjunctive to indicative prose, full of forthright assertions, providing numerous examples of well-known propositions the truth of which no leading mathematician would wish to challenge in content terms, but that remained unprovable within any axiomatic system that could capture formally the accepted laws of arithmetic. Hans Reichenbach's (1930) account of the conference proceedings was published shortly after its conclusion, but he saw no reason to mention Gödel's paper at all amongst the conference highlights. Yet his findings certainly landed with von Neumann, who wrote straightaway to say he had maybe already produced a complementary finding, inviting Gödel to meet in person to discuss how best their ideas might be developed further (Wang 1981: 654–5). This was the same von Neumann who Hilbert (1930a: 3) thought was leading the charge towards an approach to arithmetic completely divorced from the logic of quantities. However, his eventual response to Gödel's concerns (see Conclusion) was to insist instead on empirical grounding for all mathematical statements (von Neumann 1947: 196).

Gödel's incompleteness theorems extinguished all hope that any mathematical system can display absolute consistency (Reid 1963: 257). The first, as trialled at the Königsberg conference, entailed the use of statements about unprovability to show that there are essential elements of mathematics that are at once both true and unprovably so. Although expressed in high-powered metamathematical terms, it can be thought of as a variation of the riddle with which Epimenides of Cnossos both delighted and perplexed the ancient world in the sixth century BCE. With mock solemnity, he announced that his own Cretan heritage made it possible for him to say from experience that all Cretans were liars. The challenge Epimenides laid down to his audience was to decide, on the basis of the information he had presented to them, whether he himself was the liar he said all Cretans were. Two hundred years later, Eubulides of Miletus provided an even more straightforward variant of the liar's paradox: "The statement I am now making is false". Of course, the whole point of these ancient thought experiments was to establish the idea of a realm of undecidability laying between true and false. On the basis of the syntactic system which acted as the logical boundaries of these propositions, nothing at all could be decided in any definitive sense. We are deprived of any vantage point from which we can say for sure that either the proposition ϕ or its negation $\neg\phi$ is true. There consequently has to be an aspect of the truth claim in all mathematical cases of unprovability that arises from somewhere beyond the structure of formal logic that Hilbert wanted to treat as the basis of all mathematical content. Gödel's metamathematical fame arises from the fact that he was able to ground this remarkable finding in the language of number theory. If number theory, with everything we know about the basic number sequence from rote learning and intuition, is not philosophically secure, then what possibly can be?

Gödel's (2012 [1931]) second incompleteness theorem showed that there must be a requirement for some external reference point for proving that a set of axioms is consistent (Goldstern & Judah 1995: 187–8). If that set can be denoted by A, where no known logical system considers A to be anything other than consistent, can there possibly be anything amiss with the simple statement "A is consistent"? Gödel argued that there was nothing wrong with it *per se*, only that when expressed as a statement of number theory it cannot be proved from A. This is another outwardly maddening example of the existence of statements that elicit no challenge to their truthfulness in content terms but whose truth can never be proved. Mathematics is the only science which can be trusted to pass judgement on its own theoretical proofs, according to Hilbert, but Gödel countered with the striking conclusion that no mathematical theory is ever able to prove its own consistency. Every fully axiomatized mathematical system, in other words, must always permit of statements which fall short on the provability criterion within that system. Hilbert's programme attempted to remove the need for intuitions from mathematics, whereas Gödel's incompleteness theorems showed that they remained an indispensable building block of all rigorous mathematical claims (Goldstein 2005: 198). The consistency of the initial system can only be demonstrated within a stronger, yet once again fully axiomatized mathematical system. But then the problem of unprovability of consistency rears its head again, only this time requiring recourse to a still stronger fully axiomatized mathematical system. The result is a never-ending spiral of onward referral to ever stronger systems which, however ingeniously constructed, still can never overcome that final hurdle of being able to prove their own consistency.

Even mathematics, which for so long had prided itself on being the guarantor of certainty for other sciences, was thus pushed headlong into an uncertainty trap (Raatikainen 2003: 157). The dream of certain, consistent and complete foundations for mathematics, so ably and so boldly pursued by Hilbert, was recast as intellectual escapism. The second of the two theorems has normally been seen as the real hammer blow to the Hilbert programme, but within 60 years of its first airing at the Königsberg conference even the first theorem on its own was generally regarded as having stopped it in its tracks (Kreisel 1976; Smoryński 1977; Prawitz 1981; Simpson 1988). Uta Merzbach and Carl Boyer (2011: 561) say that Gödel's stipulation of undecidable propositions was as unsettling to the mathematical imagination in the early twentieth century as had been Hippasus of Metapontum's of incommensurable magnitudes two-and-a-half millennia previously. Hippasus is reputed to have drowned at sea as a punishment from the gods for having revealed – much to the frustration of the Pythagoreans of his day – the possibility of irrational numbers (Zamarovský 2017: 87). Thankfully, no such fate awaited Gödel. Instead, he received the professional respect of his peers. The public oration accompanying Harvard University's award in 1952 of an honorary degree attributed to Gödel "the most important mathematical truth to be discovered during the twentieth century" (cited in Stoeger 2011: 83).

However, there is now a revisionist historiography which depicts Gödel's incompleteness theorems as a mere "breach of the peace" (Gauthier 1994: 11), rather than something so mind-bendingly profound as to be the final word (Detlefsen 1986, 1990). There are mathematicians who have learnt to live with Gödel's findings while pressing ahead with the still rich way of thinking that Hilbert bequeathed. They have been very important to the history of economic theory. There are also mathematicians who simply ignored Gödel's incompleteness theorems even in the heyday of Hilbert scepticism. They too have been very important to the history of economic theory. The French collective known by the pseudonym Nicolas Bourbaki has been accused of proceeding with Hibert's programme as if Gödel had never existed (Mayberry 2000: 107). The most sophisticated mathematical market models within economics are still the Bourbaki-inspired general equilibrium theories of the 1950s. Mathematical economics thereby appears to be stuck at some sort of crossroads, possibly even in a metamathematical blind spot (Punzo 1991: 3). There seems to be general consent to the idea that Gödel had removed from the subject field utopian ideas that mathematics would ever reveal itself to consist solely of well-ordered structures (Israel 1981: 206). At the same time, there are very few active dissenters to the view that Hilbert remains responsible for initiating the move towards a thoroughly modern approach to mathematical formalism (Feferman 2011: 119). To understand why this tension persists, it is necessary to turn to what lies behind the "*Tische, Stühle, Bierseidel*" statement.

THE MEANING OF THE HILBERT PROGRAMME

Euclid's *Elements* begin in Book I by going straight into a series of 23 definitions of geometrical principles, from which every subsequent element of his theoretical system is derived (Euclid 1956: 153–4). The most basic principles are explained through simple drawings that capture visually how points, straight lines and planes are connected, first to one another and then to other geometrical phenomena. Given how significant Euclidean geometry was to the teaching of mathematics for the following two thousand years, it became reassuringly self-evident using visualization alone what a point, a straight line and a plane are. There was no need to have them defined from scratch every time they were appealed to, because routine experience could be used instead.

However, these are intuitive definitions that fit poorly with the axiomatic work to which they have typically been put. With a short interlude for five "postulates" and five "common notions", Euclid (1956: 155–369) completed Book I by moving straight from his 23 definitions to what he considered to be the 48 most essential "propositions" of geometry, again all expressed in words with supplementary diagrams to help the reader call to mind what is being said. In Philip Kitcher's (1981: 516–17, 1989: 432–3) terms, these are the filling instructions which helped to flesh out Euclid's general argument pattern, both here and in the *Elements'* subsequent books. However, the related proofs have been criticized by later generations of mathematicians for leading only to "foggy inferences" (Tappenden 2013: 323) that are "unsatisfactory and imprecise" (Jacobs 1992: 29). This is not to suggest that any of Euclid's subsequent 465 propositions are false *per se*, only that the route to demonstrating that they are not false is one with which mathematicians became increasingly frustrated (Dunham 1990: 38). Modern mathematicians have set themselves decisively against explanations that are reliant on "sensuous observation" (Erickson 2006: 60), doubting whether the dependence on "visual recognition" amounts to genuine explanation at all (Reid 1986: 58). "[T]he diagram contributes nothing to our understanding of the text", is André Weil's (1978: 92) rather blunt assessment.

Hilbert was by no means the first to try to remove all attribution of explanatory power to visual evidence in geometrical thinking, because Moritz Pasch and Guiseppe Peano had already struck out in this direction (Vaisman 1980: 8). Hilbert's originality lay in trying to eliminate the link between the meaning attributed to axiomatic systems and the underlying physical processes to which they referred. Economists' market models cannot be expected to have been anywhere in his mind, but the separation of mathematical models from corroborating physical evidence has cast a long shadow over them ever since. At one level, Hilbert's new approach meant substituting a series of signs linked to one another through logical reasoning (in short, proper mathematics) for Euclid's user-friendly diagrams (which, at most, could only be preliminaries before the proper mathematics began) (Corry 2008: 139). At an altogether more sophisticated level, this meant nothing short of trying to redefine the very meaning of "meaning" as a metamathematical principle.

Hilbert's 1899 classic, *Grundlagen der Geometrie* (translated into English as *Foundations of Geometry*), removed all ontological requirements from Euclid's sense of what made a point a point, a straight line a straight line, a plane a plane, and so on. They were no longer to be known for what they are shown to be under wider experimental conditions, but for how they might be represented through logic. In Hilbert's (1950 [1899]: 1, emphases in original) own words, he embarked upon "a new attempt to choose for geometry a *simple* and *complete* set of *independent* axioms and to deduce from these the most important geometrical theorems in such a manner as to bring out as clearly as possible the significance of the different groups of axioms and the scope of the conclusions to be derived from the individual axioms". Note that there is no mention at all here of the content of the axioms (Euclid's "definitions") or to what they might refer in substantive terms (Euclid's "propositions"). "Points", "straight lines" and "planes" are no longer important in themselves as points, straight lines and planes.

The system of axioms is thus liberated from the intuitionist framework of ordinary language statements about primitive objects, which means that the basic concepts related to a system of axioms can be named in whatever way is deemed appropriate, as long as the formal structure of the axioms is protected in logic (Mainzer 1996: 368). Confusing though this may seem on first hearing, it means that a point no longer needs to be called a point providing it continues to exhibit the properties that are ascribed to a point using Hilbert's logical steps. It can be called anything you want without this changing how it is understood formally as a matter of logic. Within Hilbert's logical system, a point is simply a pair of real numbers displayed in set-theoretical notation (Yandell 2010: 789). Accepting such a way of thinking, no position on a demand-and-supply diagram needs any longer to be economically derivable in any known state of the world. It can still be given an economic sounding name and appear in economic arguments as long as it is mathematically derivable as one real number on the horizontal axis and another on the vertical (see Chapter 7).

The *Grundlagen der Geometrie* was criticized for what many of Hilbert's contemporaries saw as the attempt to turn mathematics into a mere formal game that would be forever cut off from the physical processes they thought it should describe. "In the beginning", Hilbert (1996 [1922]: 1122) pronounced sometime later by way of a rejoinder, "was the sign". He thus turned on its head the German idealist philosophy encapsulated by Johann Gottlieb Fichte's (1987 [1800]: 79) dictum that "We do not act because we know, but we know because we are meant to act". The idealist conception of action translated in mathematical terms into content, but in asserting that the signs on which he relied existed prior to experience Hilbert's method of symbolic representation can seem as though it deprives mathematics of all content (Israel 1981: 216). If it can only be a purely symbolic language then the objects of mathematical analysis must be reduced to "grammatical dummies" (Tasić 2001: 71).

The question that formed the backdrop for pre-Hilbert understandings of rigour was whether a theory could be realized in a concrete domain of knowledge. A mathematical proof came to life, as it were, only when observational data fleshed out its substance relative to something that was recognisable from the physical world. Hilbert's new standard, however, asked merely whether there was an in-principle objection to the assertion that the theory would ever prove realisable (Iglowitz 2012: 333). As Volker Peckhaus (2003:

143) has observed, "Hilbert started, so to say, with the model", emphasizing the need to construct the framework in which all future explanations were to be housed before attempting any actual explanatory work. It will be seen in future chapters how often economic theorists – as well as economics imperialists – have also come to start with the model. In Hilbert's case, this meant that the test for rigour was contained within the mathematical structure itself, and that if a system of axioms could prove its own internal consistency then this was enough to say that a proof had been demonstrated. There was no longer to be deference to the idea of referential truth claims that required external corroboration, nor yet to what now looked like the old-fashioned criterion of evidence. According to Hilbert's *Beweistheorie*, a consistency proof was all that was needed (Menzler-Trott 2007: 315). A proof that survives contradictory tests can be said to admit of realization, even if it is unclear what its primitive objects refer to when translated into empirically observable objects.

The main difference between Euclid and Hilbert was not that Hilbert proved any of Euclid's 465 propositions to be untrue, only that he doubted whether Euclid had devised the best means of demonstrating the truth claims contained within them. Hilbert wished merely to dispense with reliance on the everyday definition of words for explanatory, clarificatory or even heuristic purposes. His approach to metamathematics is that *only* mathematical theories can be created about other mathematical theories (Grattan-Guinness 2000: 208). The approach to rigour initially outlined in the *Grundlagen der Geometrie* hence implies making a series of finitely many logical moves from where the axioms say the analysis should begin, but at the same time restricting both the sequencing and the nature of those moves to statements of pure logic (Sieg 2013: 14). Rigour, in this sense, relates to "a complete solution of the problem", where both the problem and its solution are to be understood only in formal terms (Hilbert 1950 [1899]: 82). Hilbert explained his position at more length in his famous Paris lecture of 1900, when he presented a list of what he believed to be the most pressing mathematical problems of his age. The "requirement of logical deduction by means of a finite number of processes is simply the requirement of rigor in reasoning". He continued by insisting that "it is an error to believe that rigor in the proof is the enemy of simplicity ... The very effort for rigor forces us to find out simpler methods of proof" (Hilbert 1902: 441).

The word "simpler" might well raise eyebrows amongst social scientists who have looked at an article written for a mathematical economics journal by the Hilbertians Gérard Debreu, Hugo Sonnenschein or Rolf Mantel (see Chapter 7). It is difficult when doing so to dispel the feeling of having been temporarily transported into a parallel universe. Of course, the standards by which one might judge the simplicity of the proof is not the way the word is used in everyday speech to suggest that something is straightforward and instinctively easy to grasp. It means instead that it has had removed all associations with the potential ambiguities that arise whenever the definitions that come from everyday speech are allowed to appear within the mathematical structure. Intuitive reasoning, along with its connections to experiences grounded in physical reality, are able to be discarded. "Simple" thus relates to relationships expressed in their most stripped-down form, where everything that is not part of the original axiomatic structure is excluded from the analysis, so as to preserve the grand edifice of purely symbolic expression.

Nonetheless, there is some dispute in the history of metamathematics literature about what Hilbert meant by his famous 1891 observation at Berlin railway station. The more literal interpretation is that he had in mind the vision of a purely content-free mathematics (Gamwell 2016: 190). This would be implied by his certainty that there must be a realm in which all mathematical problems could be resolved (Hilbert 1902: 445), whereby formal proofs will be "no different in basic, syntactic form from a finite sequence of strokes ||| ... |" (Hallett 1994: 181). However, there have been concerted attempts since Hilbert's death to rehabilitate him from some of his own more utopian accounts of the objectives that underpinned his work (e.g., Kitcher 1976; Giaquinto 1983; Feferman 1988). This Hilbert renaissance has been based on the idea that "*Tische, Stühle, Bierseidel*" was meant in only an allegorical sense. It has dialled down on just how exacting his vision for mathematics was, focusing instead on his alleged preference for model theory that stopped short of "the marks-on-paper formalism ... [of] ... the symbol-loving arithmeticians" (Grattan-Guinness 2000: 208). Hilbertian *Beweistheorie* becomes a pure "free-standing" structure under the influence of the more literal interpretation of his epiphany in the aftermath of Wiener's lecture (Iglowitz 2012: 80). But it is merely a prelude to the "exhibition of a model satisfying [the axiomatic] system" under the more allegorical interpretation (Sieg & Ravaglia 2005: 987; Sieg 1988: 338).

What the "tables, chairs, beer mugs" quote appears designed to do is draw attention to the essentially arbitrary nature of the Euclidean choice of points, straight lines and planes as the place to begin surveying the foundations of geometry (Corry 2007: 782). And a choice it was. The names given to primitive objects might usually become significant to the communicative practice of illustrating the principles in question, but it is mathematically insignificant to the derivation of the principle itself. As Constance Reid (1996: 60, emphases in original) explains, whatever the primitive objects of an axiomatic system are ultimately called, "they would *be* those objects for which the relationships expressed by the axioms were true. In a way", she continues, "this was rather like saying that the meaning of an unknown word becomes increasingly clear as it appears in various contexts". There is a purging process in operation here, leaving only mathematical exposition denuded of direct physical resemblance.

Gödel's incompleteness theorems suggested that some role had to be left open for intuition even in Hilbert's preferred formal axiomatic systems. This much was fully in the public domain by 1931, which immediately places Blumenthal's recollection from 1935 of the "tables, chairs, beer mugs" quote in a rather different light. Indeed, already by as early as 1905, Hilbert was in print suggesting that the essential arbitrariness of the foundations of geometry was such that the move from there to the foundations of arithmetic was likely to prove tricky (Hilbert 1905: 338). One might just as well start with circles and spheres, he said, as points, straight lines and planes. This would imme-diately have seemed more familiar in Euclidean terms than if "*Tische, Stühle, Bierseidel*" literally meant "tables, chairs, beer mugs". However, what Hilbert appears to have had in mind is that the names of the primitive objects are so unimportant to what followed that the act of naming could be entrusted to a random letter generating machine. Hilbert wanted geometry to become a formal deductive system (Jagnon 2006: 67), and to achieve that then the names of concepts needed to be separated from the operation of the deductive struc-ture (Hilbert & Cohn-Vossen 1999 [1932]: 66). The primitives had to remain undefined while the purely mathematical processes took place on the axioms (Kline 1972: 1010). The axioms driving the mathematical practice were thus themselves to be written down solely by using the symbols of formal logic and without the potentially corrupting influence of words (Breger 2000: 228). Words, of course, are also generally absent today in the proofs to be found in the leading economics journals.

Primitive objects are to be treated as "bare 'things'" that can be assumed to come to life only after the satisfaction of the system's axioms have distinguished between those theorems that are true and those theorems that are false (Rodin 2014: 40). They remain undefined terms with no essential meaning throughout the task of proving the theorems. "*Tische, Stühle, Bierseidel*" are mere letter sequences for the duration of this process, as are "points, straight lines, planes". It is only when all proofs are at hand that meaning is then assigned to the concepts of a system that is already known to be true in its own terms (Sentilles 1975: 93). At this point it is almost certain that "*Tische, Stühle, Bierseidel*" will not mean "tables, chairs, beer mugs" at all. In mathematical economics, then, does "demand and supply" in a Hilbertian sense still mean what we would typically associate with demand-and-supply dynamics as they relate to the world of real phenomena? What about "consumers, producers and workers"? To get to the heart of the relevant differences, further reflections are necessary on exactly what Hilbert was trying to achieve for mathematics.

HILBERT IN THE HISTORY OF MATHEMATICAL RIGOUR

Today, Hilbert is irreducibly linked to the rigour movement in mathematics. This is a broad trend going all the way back to the fifth-century BCE Pythagoreans and their desire to embed the very nature of mathematical thinking within the concept of proof (Corry 2004: 5). However, acceptance of the need for proof only followed the prior acceptance of the process of experimentation (Elkana 1986: 64). This first move occurred in late antiquity, around eight centuries after Euclid laid down his *Elements* (Beller 2001: 229). The debate about mathematical rigour has historically revolved around where exactly one should be situated between the twin poles of proof-making and experimentation, between arguing that mathematical proofs are standalone artefacts exhibiting their own truths and demanding that their claims are suitably grounded in physical reality (see Chapter 3). The Hilbert of Berlin railway station seems to be positioned squarely at the proof-making pole, but a more rounded assessment of his overall contribution reveals considerable ambiguity in this regard.

Proof was established from very early on as that which goes beyond a mere example of an interesting phenomenon to capture the essential characteristics of that phenomenon in generalized form. This is what the historian of mathematics Leo Corry (2008: 130) has called "the regulatory idea of proof", that special something which appears to offer access to a greater truth than can be established through *ad hoc* claims based on intuitive appeals to numbers and diagrams. The commitment to proof has remained intact for at least two-and-a-half thousand years, but it is group dynamics amongst the community of practicing mathematicians which then determines what *type* of proof is considered most desirable. In Philip Davis and Reuben Hersh's (1986: 66) somewhat stark terms, what subsequent generations of mathematicians learn as a proof only "testifies that the author has convinced himself and his friends that certain 'results' are true". Corry (1993: 111) writes of a scientific revolution of genuinely Kuhnian proportions that has endured since the time of classical Greece. To go in any direction other than that implied by the "deductive transformation" would not now conform to the idea of doing mathematics, or at least not to doing it properly. Yet this still does not tell us to what degree the commitment to deductive proof will be accompanied by a secondary commitment to demonstrating the practical purchase of the proof through appeal to experimental data.

Foundational debates concerning the calculus are highly instructive in this regard. Chapter 1 has already shown the significance of Lagrange's reworking of the differential calculus for the related concept of explanatory unification and, from there, for contemporary practices of economics imperialism. But what do we learn from the earliest ontological objections to the calculus? The first sustained critique of Isaac Newton's and Gottfried Leibniz's separate renditions of the calculus came from the Irish philosopher George Berkeley in the 1730s (Berkeley 1754 [1734]: 28). He had previously exonerated them from the charge he laid against other mathematicians of being "meer triflers, meer nihilarians" (Berkeley 1871 [1707–08]: 497). Yet still the calculus was placed in his sights because of the threat to his theological beliefs he saw within it, in particular its explicit flirtation with the idea of infinity. The space in which the calculus prospered "exists necessarily of its own nature", Berkeley (1992 [1721]: 98) complained; "it is eternal, uncreated, and even shares the divine attributes". Mathematicians had to be treated with suspicion, he concluded, because they were forever dividing appearance from reality and retreating into the realm of the imperceptible. Berkeley's philosophy was grounded in a theory of perception (Ardley 1968: 35), and mathematics became theologically suspect if it departed to any degree from perceptible quantities of finite extensions (Breidert 2005: 504).

Alongside this theological critique, there were also more obviously mathematical concerns (Guicciardini 1989: 38–41). Berkeley believed that the calculus represented all that was wrong with the developing tendency to allow mathematical discourse to focus on the derivation of one set of arbitrarily chosen signs from another by following pre-set procedural rules (Jesseph 1993: 117). Purely nominal elements of mathematics could only be justified, Berkeley argued, if they led directly to practical insights that would assist in the process of curing everyday problems. Anything else, he continued, would serve only to illustrate how far mathematics fell short of genuinely rigorous standards (Kalman 2009: 214). His suspicions here are not in relation to deductive proof-making *per se*, so much as to the failure to stipulate how those proofs might come to life through being harnessed to the experimentation that had helped to define the mathematical process since late antiquity. Berkeley's charge remained unanswered for almost a century and a half, as the mathematical tools for developing a genuinely rigorous basis for the calculus were simply not available, in spite of prodigious efforts made towards that end by Jean le Rond d'Alembert, Joseph-Louis Lagrange and Siméon Poisson (Agassi 2008: 47). As Judith Grabiner (1983: 189) has written, the key was to be found in understanding that "an equation involving limits is a shorthand expression for a sequence of inequalities". Even the mathematician eventually delivering the proof, Augustin-Louis Cauchy (2009 [1821]: 21–48) in his *Cours d'Analyse* of 1821 (in English, *Analysis Course*), did so using a purely verbal definition of limits which sat outside the algebraic frame in which it was to be used.

The convoluted process by which rigorous foundations for the calculus were eventually demonstrated merely emphasized the different options available between the twin poles of proof-making and experimentation. On the one hand, there was the mid-eighteenth-century French mathematician the Comte de Buffon, who rejected any sense that there could be a structure of autonomous proof-making separate from real-world calculations (Roger 1997: 38). Mathematical proofs might always veer towards sterile syllogism in Buffon's view, capable of confirming nothing other than the logic of their own starting point (Reill 2005: 40). He frequently bemoaned the tendency towards "emptiness" within pure mathematics, worrying that this might cut adrift the whole of the subject field from factually grounded investigations into the nature of the world (Grabiner 1981: 313). The shadow of Berkeley thereby persisted. On the other hand, there was the late eighteenth-century Italian/French mathematician Joseph-Louis Lagrange, who denied the very validity of Berkeley's opening premise (Gaukroger 2010: 143). He believed that if it was to be possible to determine the rigorous basis of the calculus then what was required was for it to be formulated in strictly algebraic terms, rather than trying to copy what could be seen in the physical world (Bos 1980: 90–91). In which direction would Cauchy (2009 [1821]) jump?

His so-called delta-epsilon proof for the calculus would appear to echo the experimental "emptiness" of which Buffon complained and position Cauchy much closer to Lagrange's proof-making techniques (Grabiner 2010: 83). As with so many other issues with which this book is concerned, though, first impressions can mislead. Cauchy certainly seems to have set off with the same intentions as Lagrange, in Joan Richards's (2006: 712) words to deal decisively with the "confusions" that had for so long stymied the search for rigorous foundations. Yet they went about their shared endeavour in different ways. Lagrange's plan of attack was to say that if mathematical concepts such as the limit and the infinitely small were to be used – concepts that were fundamental to the mathematical insights delivered by the calculus – then they should be formally derived (Jahnke 2003: 107). Cauchy's aim was instead to find precise and operational definitions for those concepts (Laugwitz 2000: 181). Their disagreement is therefore easily stated. Lagrange wanted the narrative elements of the calculus to disappear altogether, but Cauchy – whose original ground-breaking solution to Berkeley's challenge, remember, was presented in purely verbal form – only wanted to exclude the narrative uncertainty that followed from not securing definitional clarity. What might have looked from Buffon's perspective like fully empty mathematics in Lagrange's case was therefore not necessarily so in Cauchy's.

Hilbert's formalism promoted a self-standing apparatus for mathematics, a discipline free of intuitions, free of narrative elements and free of the need for external validation. But Cauchian metamathematics appears to lead elsewhere. Gustav Dirichlet dreamt in explicitly anti-Cauchian terms, then, when proposing a mathematics where it was necessary "to put thoughts in the place of calculations" (Ferreiròs 2007a: 244). Hilbert positioned himself at the end of the lineage that included Dirichlet and Richard Dedekind, but was therefore progressing down only one road of many with his notion of arithmetical sentences without quantifiers (Gauthier 1994: 2). Dirichlet, it can be presumed, would have approved of the fact that Hilbert's system of axioms was populated by "thought things", or the product of his own capacIty to imagine them into existence. In what has become a famous passage from a 1905 lecture, Hilbert said: "I have the ability to think *things*, and to designate them by simple signs (*a, b, … X, Y, …*) in such a perfectly characteristic way that I can always recognize them again without doubt". It was this lack of ambiguity which allowed him to believe that, within a perfectly specified system of axioms, "points, straight lines, planes" could prove formally interchangeable with "tables, chairs, beer mugs". He continued: "My thinking operates with these designated things in certain ways, according to certain laws, and I am able to recognize these laws through self-observation, and to describe them perfectly" (Hilbert 1905, cited in Peckhaus 2003: 150). There is an echo of Hilbert's later argument in James Clerk Maxwell's (1890 [1870]: 216) allusion in 1870 to the process through which "Thought weds Facts" (see Chapter 1), an argument that half a century later Hilbert (1966 [1922]: 1122) had reduced to the much simpler motto of "In the beginning was the sign".

Yet this was Hilbert continuing to promote his programme in the knowledge that there was significant metamathematical resistance to it. In the 1880s, the German logician Gottlob Frege (1950 [1884]) had published an important treatise which sought to discredit the view that mathematical objects owe their existence solely to how the minds of mathematicians function (Goodman 1991: 120). Frege appealed to the authority of Joseph Fourier to make his case, where Fourier (2007 [1822]) had claimed that free-standing mathematical structures had no intrinsic meaning in their own right (Kline 1953: 287). Instead, meaning is ascribed to them when concrete numerical content is added through the calculations that are enabled by experimentation. Fourier emphasized the intimacy of the relationship between mathematics and the study of nature, and Cauchy appears to have been a dedicated follower in this regard.

Hilbert (1998 [1897]: x) summed up his own approach early in his career while inadvertently also highlighting the continued presence of competing metamathematical camps. "I have tried to avoid Kummer's elaborate computational machinery", he explained, "so that here too Riemann's principle may be realized and the proofs completed not by calculations but purely by ideas". This set him apart from those who believed that mathematics should be direct and computational (Fré 2018: 152), and perhaps even further apart from those who believed that it was to step outside the mathematical tradition altogether to behave in any way other than this (Stedall 2012: 28). Henri Lebesgue had argued at the turn of the century that there might even be something some-what deceitful in formulating *as* fact statements about mathematical relations that it was impossible to demonstrate *in* fact (Bressoud 2008: 127). Logic was logic and fact was fact, he said, and the two cannot be treated interchange-ably. Cauchy (2009 [1821]: 2) stated his intention to "make all uncertainty disappear, so that the different formulas present nothing but relations among real quantities, relations which will always be easy to verify by substituting numbers for the quantities themselves". Hilbert's willingness to treat state-ments cast in logic as statements of fact suggests that he was working to a very different conception of mathematical truth to Cauchy.

A contemporary of Hilbert's, Jacques Hadamard (1996 [1905]: 1084), wrote revealingly about the difference between functions that are "defined" and functions that are "described". The former encompasses truth claims in which it is possible to accept statements of logic as statements of fact, but the latter does not. Defined functions are propositions regarding existence, related to the hypothetical possibility that an entity is findable rather than to evidence that it has actually been found. Described functions instead need to demon-strate evidence of the material properties of that entity (Otte 2007: 253). In Kantian terms the difference is between analytic sentences, deemed true on the basis of the meaning of the terms involved in the sentence, and synthetic sentences, deemed true on the basis of external empirical validation of the matter to which the sentence speaks (Goodman 1991: 120). The scene is thus set to contrast Hilbert, figurehead of the new generation, as the proponent of defined truths founded on analytic mathematical sentences with Cauchy, figurehead of the old guard, as the proponent of described truths founded on synthetic mathematical sentences. Once again, though, dangers lurk in rush-ing to judgement, because the Hilbert of Berlin railway station fame differed in important ways from Hilbert in later life.

The evidence from Hilbert's more mature work is much less suggestive of a desire to turn mathematics into a content-free practice revolving solely around the manipulation of symbols for its own sake (Tomalin 2006: 44). He increasingly asked two key questions in the research that focused the discussion directly on substantive matters. The first was *how* the content should be brought in: Hilbert's preference for formal axiomatic systems meant not through ordinary language statements that invited disputes over the definition of words. The second question was *when* the move to something more concrete should occur: Hilbert's preference for formal axiomatic systems meant not until all the axioms had been thoroughly worked through in purely mathematical terms. In an attempt to clarify these issues, he distinguished in his later *Grundlagen der Mathematik* (*Foundations of Mathematics*) between the separate processes of concrete axiomatization and formal axiomatization (Hilbert & Bernays 2003 [1934]: 3–4, 10, 18, 25). Neither was to be considered right or wrong in its own terms, only that they needed to occupy their own particular place in the mathematical imagination. (There are echoes here of Jevons's (1876: 624) understanding of a similar internal division of labour within economic theory (see Chapter 4)). Important qualifications to Hilbert's formalism therefore abound even in his own work.

Chapter 1 of Volume 1 of the *Grundlagen der Mathematik* addresses the content issue head-on. In collaboration with Paul Bernays, Hilbert describes a series of *"inhaltliche"* systems that have clearly enriched mathematical knowledge. Their list includes Euclid's geometry, Newton's mechanics and Clausius's thermodynamics (Hilbert & Bernays 2003 [1934]: 2). Following Stefan Bauer-Mengelberg's usage (see Hilbert 1967 [1925]: 376), *"inhaltliche"* is generally translated today as "contentual", and it refers to those processes of concrete axiomatization when Hilbert's rules for purely formal axiomatic procedures are not followed. Here, the primitive objects used in the investigation have content ascribed to them from the start, where that content is drawn from directly describable experiences (Sieg & Ravaglia 2005: 987; Rodin 2014: 68). It is when discussing cases of this nature that Hilbert's work looks most like Cauchy's. Newton's observations of the apple falling from the tree represent a perfect example of how content initially precedes the mathematics in *"inhaltliche"* exercises. The trigger for his mathematical explanation of gravitational forces relies upon an event impacting upon his intuitive reasoning in an entirely unpredictable manner. Newton's apple did not fall in a carefully conceived experimental setting, but it could have done so in a way which brought together the two classical second-order concepts of proof-making and experimentation in a pure *"inhaltliche"* moment.

Such moments, however, can never begin to approach the purest forms of thought experiment that the younger Hilbert (1902: 438) held so dear. There was too much of a fundamentally *ad hoc* nature here for mathematics ever to be grounded in "absolute certainty" for the Hilbert who wanted to assert the fundamental likeness between "points, straight lines, planes" and "tables, chairs, beer mugs" in a logically sound axiomatic system (Moore 1982: 255). The later Hilbert remained anchored at least partially in a world of *"inhaltiche"* moments, as demonstrated by his turn away from studying the axiomatic principles of mathematics to studying the axiomatic principles of physics (Gray 2000: 185). Taking his work as a whole, he can be seen as having one foot in two camps. The distinction that he and Bernays drew between concrete and formal axiomatization therefore represents more than merely two metamathematical paths open to mathematicians. They were the two paths that he trod often simultaneously during his own career.

Hilbert's advocacy of formal axiomatization still almost certainly remains the work for which he is best known, even if it reflects only part of his overall corpus (Ryckman 2016: 23). It speaks to a model world where, in his own words, "we simply have concrete signs as objects, we operate with them, and we make contentual statements about them" (Hilbert 1996 [1922]: 1123). The *"inhaltliche"* moment does not disappear in this version of how to proceed metamathematically, but note the sequencing has been reversed. For formal axiomatization, the ascription of content to the mathematical relations no longer precedes how those relations are constructed, but comes after their internal logic has already been fully specified. This is working with mathematical functions that have been defined into a particular role within the axiomatic structure rather than having been described by the role they play within the real world. It encapsulates a vision of the research process which admits of the possibility of all mathematical problems being solved for all time as long as the right rules are instituted (Breger 2000: 228).

To follow Hilbert in this endeavour means restricting the analysis on the assumption of a fixed system consisting of a "previously *delimited domain of subjects* for all predicates from which the statements of the theory are constituted" (Hilbert & Bernays 2003 [1934]: 2, emphases in original). There is no claim that this represents anything other than an "idealizing assumption", but it is justified within the demands of a formal axiomatic system because it, and *only* it, succeeds in "joining the assumptions formulated in the axioms" to provide for a tractable mathematical model (Hilbert & Bernays 2003 [1934]: 3). Hilbert clearly had a lot of respect for those who engaged in the process of concrete axiomatization, but thought that this was most appropriate when a science was still taking shape and mathematics was obviously already well beyond that stage. It remained necessary, he argued, for less mature sciences to start with a body of empirical knowledge that is already widely accepted by its community of practitioners, but that mathematics was ready for something else that might deliver the holy grail of complete certainty in foundational knowledge of itself. Presumably both mathematical economics and the economics imperialism it has spawned look like highly embryonic sciences on this typology, irrespective of what their most insistent proponents might say.

The image that emerges is thus a very complicated one. However, it is less important exactly who fits where in this grand metamathematical debate than it is to recognize that disagreement has always raged over the relative importance of proof-making and experimentation. Buffon, Fourier, Frege, Lesbegue and Kummer are undoubtedly amongst the great names of mathematics; so too are Lagrange, Cauchy, Dirichlet, Dedekind, Riemann and Hilbert. Yet there is no single position that can encapsulate their metamathematical commitments. Even mathematics, by common consent the most unified of disciplines, resists straightforward homogenization in metatheoretical terms. However, if there is no single mathematics, there can be no single mathematical economics and no single approach to the mathematization of economists' market models. It consequently makes little sense to think in terms of "better" models displacing those that have self-evidently had their day or of a necessary progression towards ever higher learning. As a participant in metamathematical debates himself, it is interesting to note that Hadamard said in 1905, at the height of Hilbert's search for purely formalist foundations for mathematics, that "there are two conceptions of mathematics, two mentalities, in evidence" (Hadamard 1996 [1905]: 1084). This phrase can be read as a weary sigh that his own view was destined never to fully win out or as a celebration of the resulting pluralism of mathematics. Either way, we should treat it as an important reminder that there will always be multiple ways of conducting mathematically oriented research in economics as well as multiple routes to the mathematization of economists' market models, and that all of this multiplication should be expected to exist side by side in complex ways in the thoughts of economists. It will also be joined by those who operate on a different ontological plane altogether and resist the lure of mathematical explanations in preference for the accumulation of in-depth empirical knowledge. The prehistory of economics imperialism to which Chapters 4–7 will turn is therefore unashamedly a patchwork history of different moments which have each left their separate imprint on economists' contemporary practices. It can never be a simple linear history.

CONCLUSION

Hilbert struck both awe and fear into his contemporaries for his ability to propose a single "thought-thing" as a substitute for the potentially endless industry of calculation. When Paul Gordan, the King of Invariants to the mathematicians of his day, first saw how his prodigious calculative efforts could be so neatly sidestepped by a three-page proof (see Hilbert 1890, 1893), he remarked in clearly reverential tones: "This is not mathematics, it is theology" (cited in Weyl 1944: 622). Hilbert latterly appropriated Gordan's only ever semi-serious distinction to revisit some of his earlier metamathematical work in which he insisted that all mathematical theories are likely to pass through three phases on their road to maturity (Corry 2004: 144; Thiele 2005: 251). He distinguished between a theory's naïve phase (governed by intuitive statements relating to the relationship between different mathematical entities), its formal phase (triggered by the move to symbolic calculation as a replacement for intuition), and its critical phase (when the purely logical properties of the underlying mathematical relationship could finally be spelt out in full). Hilbert understood his own embrace of defined functions, wherein mathematics increasingly revolved around the production of existence proofs, as a necessary step towards critical metamathematics.

However, there is a curious chronology here. It was the early Hilbert who invested unquestioned faith in the possibility that mathematics could become the accumulation of ever more general existence proofs, but it was the later Hilbert who used Gordan's theology quotation – 25 years after the fact and out of context – to illustrate the resistance that his proof theory was still experiencing. Yet the later Hilbert was no longer conducting his own mathematical research in a manner that would have made most sense to his younger self.

His work can neatly be packaged into a series of largely discrete stages, and by the late 1920s he had other methodological targets in mind beyond Gordan's commitment to comprehensive calculation. Indeed, in his efforts to bring his own mathematical insights to the foundations of physics, he appears to have moved back from the critical to the formal phase of theoretical construction to work with methods he had once decried in others (Webb 1997: 2).

This shows how difficult it is to provide a picture of a single Hilbert making a single argument for a single mathematics. Economists' mathematical market models must therefore also embody similar complexities. Complete unification is such an exacting objective that we should never expect the elimination of alternative paths emerging out of the same starting point. It is possible for mathematics to be at once the most unified subject field and nonetheless still far from completely unified. Internal dissent and divergent trajectories will always be the order of the day. Hilbert's career illustrates this point perfectly. He took on different sparring partners at different times, and he appears to have inhabited different metamathematical commitments accordingly. "The Hilbert programme" is a phrase that always seems to be used in the singular, but there are important differences to be detected in his evolving approach pre- and post-Gordan and pre- and post-Gödel. Traces of various metamathematical approaches are evident throughout his work. What does this mean for both mathematical economics and economics imperialism?

As Chapters 4–7 seek to demonstrate, each time a metatheoretical move was made in economics that provided another foundational element for economics imperialism, this took place against the backdrop of attempts to resolve the tension between mathematics by definition and mathematics by description. The type of discussion that appears in this chapter might well have been forced a long way into the background when these crucial innovations in economic theory occurred, but once you begin to look for its key features their imprints become increasingly hard to miss. The distinction between defined and described mathematical functions is my analytical starting point in all that follows. It informs the discussion in the following chapter of economists' practice of hypothetical mathematical modelling, allowing me to ask what sort of mathematical objects, exactly, are the market models that propel economics imperialists' uninvited border transgressions. This chapter has laid the groundwork for what comes next by reviewing what leading turn-of-the-twentieth-century mathematical theorists said about the relationship between the world of their mathematical models and the world of direct experience. Even the grandest and ostensibly the most epistemologically compelling declarations that free-standing mathematical models can assert their own truths have flattered to deceive. It is crucial for everything still to come that not once have they registered a knockout blow that has silenced those with legitimate metamathematical reservations about banishing into exile the world beyond the model.

CHAPTER 3

The autonomy of the world within the model: philosophical reflections on hypothetical mathematical modelling in economics

INTRODUCTION

It is possible to think in a manner that has a clearly mathematical underlying rationale but without surrendering to overtly mathematical expression. However, this is not a path that economists, in general, chose to take. The previous chapter reviewed key developments in the process of argument through mathematical postulation, where mathematical relationships only have to be findable in principle. Mathematical truths could henceforth be asserted through proof-making without the need for external validation of the conditions under which such truths were likely to manifest themselves in practice. Even though this never became consensus metamathematical opinion in the manner of a Kuhnian paradigm, enough mathematical economists were convinced by this core ontological claim for it to have left a lasting mark on economic theory from the 1950s onwards. This chapter focuses on the parallel turn towards hypothetical mathematical modelling as a second means of translating mathematical instincts into direct mathematical expression, where this time we are more likely to be told that the relevant mathematical relationships have actually been found. The key factor here is the similarity between what is causing equilibrium in the model world and the closest corresponding causal mechanisms in the real world. What might be demonstrated as being true in the model is also on many occasions treated as being true of the world beyond, even if the explanation of why such resemblance holds is often somewhat sketchy. We are returned to the fundamental ontological difference between a model world that might be thought into existence in its own terms and what lies beyond these hypothetical relations in real-world economic experiences. The distinction between defined and described mathematical functions simmers away just below the surface.

The broader discussion about how realistic model-world relationships have to be receives precious little attention from economists, and the distinction between defined and described mathematical functions none at all. It is therefore unsurprising that the dividing line between mathematical postulation and hypothetical mathematical modelling has become increasingly fuzzy over time. Léon Walras's 1874 classic, *Éléments d'Économie Politique Pure* (usually translated into English as *Elements of Pure Economics*), was very clearly a theoretical exercise in mathematical postulation (Walras 1984/1954 [1874]). Apart from when working through his economic definitions or pausing to criticize his non-mathematically oriented predecessors, he reasoned only with mathematical objects: be they numerical examples, diagrammatic representations or, in the main, systems of equations. Over 400 pages of text are constructed in this manner, a thought experiment of what an abstract economy would have to look like were it to exhibit the features of a competitive, price-taking, equilibrium economy. The mathematics on display were rudimentary by mathematicians' standards but wholly revolutionary as a mode of reasoning for the economics of Walras's day. Less than 20 years later, Irving Fisher presented the plans for building a hydraulic model of a three-person, three-commodity economy, to be constructed using balances, cams, levers, cisterns and valves, in an attempt to learn more about the dynamics that might produce the same conditions in practice as Walras was able to write down solely through his mathematical postulation. Fisher's mathematics were very similar to Walras's, in fact so similar that he was at pains to explain that he had "read no mathematical economist except Jevons" when concluding his own initial studies (Fisher 1892: 4). The mathematical expression in Fisher's work initially brought him extremely close to Walras, but he treated this merely as a preparatory stage for building a functioning physical model of that same economic system. Both were initially attempting argument through mathematical postulation, with Fisher alone adding model-based argumentation as a further activity, with the distinction between the two always clear.

What, though, about Kenneth Arrow and Gérard Debreu's definitive demonstration of the conditions for general equilibrium in their rightly feted 1954 *Econometrica* paper, "Existence of an Equilibrium for a Competitive Economy"? Its proof-making objectives show that it is evidently economic theory via mathematical postulation, but has it not been used primarily as a tool for learning how economic theory relates to the world beyond, just as a model would be? And what about Robert Lucas's famous 1972 *Journal of Economic Theory* paper, "Expectations and the Neutrality of Money"? Or Finn Kydland and Edward Prescott's equally famous 1982 *Econometrica* paper, "Time to Build and Aggregate Fluctuations"? What are we looking at here? Is it economic theory arrived at through mathematical postulation or the development of an analogue economic system arrived at through hypothetical mathematical modelling? The distinction between the two has recently become so blurred that it would seem impossible to say. Extended refinements in the mathematization of economists' market models have taken place simultaneously through two distinct modes of reasoning, but the point at which mathematical postulation stops and hypothetical mathematical modelling starts is much less distinct than it once was. It is often far from obvious whether what we are looking at is economic theory produced via mathematical proof-making or an economic model constructed as an object of a self-made hypothetical world.

The fact that there is no easy way of distinguishing between these two forms of mathematical expression might not be a problem except that they operate to distinct ontological standards (Mäki 2009b: 39). Economic theory through mathematical postulation entails only phenomenal existence. The abstract economic system being described in such proofs will never be mistaken for the actual economy of lived experience. However, economic models are required to act as more than straightforward mathematical postulation because, in the conventional account of their epistemic function in the philosophy of science literature, they observe a representational relationship with something that can pass as real (Cartwright 2007: 234). Something beyond mere phenomenal existence is being invoked. This should place much stricter constraints on the economic content of the assumptions that enable the tractability of hypothetical mathematical models than it does on the assumptions that drive theoretical advances by mathematical proof-making.

Economic models speak across two distinct worlds, one phenomenal and one real, what Mary Morgan (2012: 30–37) calls the world within the model and the world that the model is taken to represent. The purely mathematical nature of the structure of the world within the model means that the algebraic descriptions of its internal relations can equally be said, in line with Hilbertian metamathematics, to be true by definition (see Chapter 2). But what economist would ever want to admit to having limited themselves to truth claims relating only to the phenomenal realm of their self-made model worlds? Surely the whole point of the often extreme effort they expend in modelling different hypothetical economic relationships is that this provides them with a standpoint from which to comment authoritatively on real-world dynamics? How might the condition of being true in the model also relate to being true in the world beyond? On the whole, economists have been notoriously cagey on this question (Rubinstein 2006: 870; Grüne-Yanoff 2009: 86). It is as if we are being encouraged to think that there must be some link between model and target, but without being told what this link could be. The audience for economic models would therefore seem to be split in two. On the one hand, we have those already in the know, for whom it can be taken as a professional article of faith that hypothetical mathematical models resemble in some way the world in which we actually live. On the other hand, we have those who are desperate to be told exactly how this relationship is deemed to work.

It is an oversimplification to suggest that this division reduces neatly to economists on the side of the believers and everyone else demanding more evidence before they will join them there. It was, after all, an economist, Robert Sugden (2000: 3), who first brought to the philosophy of science literature on explanation the concern that his professional colleagues were being something other than transparent about the relationship between their models and the world beyond. He suggests that economists tend to lead the unsuspecting reader on something of a dance. Their analyses will almost always start by presenting the outline of a real-world problem related to real-world markets, with an apparent promise that the ensuing model will allow for inferences as to how that problem might be solved. Yet at most the proffered solution is likely to be contained within the hypothetical world created for the purpose of the modelling activity, with no evidence provided for why we should accept that the world beyond the model responds in the same way as the world within the model. According to Sugden (2000: 5, 2009a: 4), the real-world markets that provide the backdrop for saying that the study is of obvious real-world consequence typically serve no further purpose once they have been used to establish the reason for constructing the hypothetical mathematical model in the first place. They are central to the justification of why the model is needed, but they do not reappear in descriptions of what the model allows the reader to conclude.

There is obviously a lot to unpack here. It is the mathematical properties of market models that economics imperialists celebrate for the extra rigour they can bring to those areas of the social sciences in which they have so far made least headway. However, nobody seems to be able to say for sure how the mathematization of market models links those things that are true in the model with those things that might be true of the world beyond.

In an attempt to shed further light on these questions, the chapter now proceeds in three stages. In section one, I review the orthodox historical account of how economics has been transformed from a verbal to a modelling science, as a way of asking what types of models have come to dominate. The philosophical literature suggests that so-called surrogate systems offer relatively straightforward means of testing whether true in the model conforms to verifiable truth claims in the world beyond. However, most economic models – including economics imperialists' market models – are not of this nature. Instead, they are substitute systems, where the hypothetical mathematical model stands at a crucial distance from such testing. In section two, I ask how explanation might proceed from a mathematical market model in circumstances in which inductive inference appears to operate from a substitute system to a target that has no presence within that system. This seems to suggest that the motion the model economy describes is activated by the content of the assumptions on which it is constructed, not on empirically derived knowledge of how the world actually works. In section three, I focus on how substitute systems might be treated as recognisably economic models if they exist in a separate domain to all known economic worlds. Economics may well have progressed from a verbal to a modelling science, but its models nonetheless possess crucial narrative content. They are persuasive as autonomous investigative tools insofar as the storylines in which they are embedded carry commonly accepted theoretical principles. Mathematical market models have the potential to become something more than mathematical postulation only in the context of an audience that will accept as real their accompanying market-bound storylines. The relatively easy comeback from those seeking to resist the advances of economics imperialists is that they have already compiled ample counterevidence to these narratives. Simply imposing the market model template onto non-market phenomena will never close this ontological gap.

THE RISE OF ECONOMICS AS A MODELLING SCIENCE

Prior to the initial forays into marginalism in the mid-nineteenth century, economic theory tended to occupy an interesting half-way house. It was imprinted with a structure of thought that seemed to evoke mathematical instincts, but it was presented to the outside world through a mode of reasoning that was almost always explicitly anti-mathematical. It brought the important features of the economic world to life through a discussion of ratios, percentages, proportions, quotients and fractions, set against the theoretical backdrop of the elucidation of trends, the identification of tendencies and the characterization of stability conditions. The logic of number dominated. However, the mode of reasoning was not itself condensed into algebraic forms in anything other than a few isolated instances. In general, it belonged to the avowedly Cauchian world of the mathematics of its time. Numbers were always a real category used for illustrating the quantities of equally real economic phenomena. For three-quarters of a century following the marginalist revolution of the 1870s, most economists still used a knowingly literary style to pursue their arguments (Andvig 1991: 450; Poovey 2008: 2). This covers the first two of my four stages of the prehistory of economics imperialism (Chapters 4 and 5).

For anyone who is used to reading only the most up-to-date contributions to economic theory, it can come as a surprise just how long the verbal mode of reasoning persisted. It is not that words have completely disappeared from articles in economic theory today, merely that they have been handed a very different job. The literary genre that typified early interventions in economic theory used words to capture situations that readers would recognize from their own lives, before then constructing familiar mental imagery to explain the economic principles that were in play in those situations. Now, though, the words that feature in theoretical economics stand one place removed from the mode of reasoning, being reduced to a secondary position behind mathematical content. Specifically economic-sounding words are used to describe the structure of the hypothetical mathematical model so that it takes on the appearance of something more than an endless script of algebraic notation. Yet the reasoning itself is all contained within the choice of mathematical structure.

The shift from a literary to a mathematical mode of reasoning was an incremental stop-start process. It is no more reducible to a single moment of time than was the closely related birth of modern hypothetical mathematical modelling. Jan Tinbergen (1935: 370) was the first to talk explicitly about economists' modelling endeavours – *"Zu diesem Zweck haben wir nun ein Modell der Wirtschaft zu konstruieren"*, "To this end, we now have to construct a model of the economy" – when building what he called a simplified model with its accompanying equations. This was in the mid-1930s. He was awarded the first Nobel Prize in Economics, jointly with Ragnar Frisch, in 1969. The citation from the Nobel Committee (1969) said that their award was "for having developed and applied dynamic economic models for the analysis of economic processes". There is, then, a roughly 35-year period between the initial transfer of the label "model" from physics to economics and the two most important pioneers receiving the ultimate professional accolade for their modelling endeavours. This period therefore bookends the move to a mathematical mode of reasoning. Even here, though, things are not necessarily straightforward, as the class of models they developed differed in important ways (Boumans 2005: 14). Tinbergen (1937) created statistical objects, Frisch (1933) algebraic objects. Mathematization in economics has always had more dimensions than simply adding numbers to existing theoretical propositions.

The subjects of my historical chapters often sit uncomfortably in relation to these bookends. Jevons (the subject of Chapter 4) was working at a time preceding the wholesale embrace of hypothetical mathematical modelling. An embryonic mathematical mode of reasoning is evident throughout his work, but his account of theoretical principles relies every bit as much on verbal as on mathematical reasoning. Compared to most of what had come before, his principles text of 1871 does look different. His account of the strictly limited historical precursors to his approach, which he included in the preface to the 1879 edition, reinforced the sense of striking out into largely unknown territory. But *how* different exactly? Some elements of his work are clearly a refutation of the classical tradition on which he was building, but others are equally clearly a direct continuation. The fact that he was content to reason verbally around examples that displayed his mathematical instincts is one instance in which everything seemed much the same. Robbins (Chapter 5) operated in the same vein. He worked with theoretical principles that seemed to make most sense as mathematical statements, but always reasoned verbally. His career was roughly coterminous with the period in which the shift to hypothetical mathematical modelling occurred, but that shift left no discernible mark on the content of his theoretical outputs. Indeed, he seemed to become ever more the mathematical sceptic as he aged. It is only with Samuelson (Chapter 6) and Arrow and Debreu (Chapter 7) that we see the mathematical mode of reasoning truly flourish and theoretical commitments come to life within hypothetical mathematical models. Yet despite sharing a preference for a modelling methodology, there are palpable disagreements between them about the very form that an economic model should take.

The philosophical literature on the nature of explanation now routinely refers to the distinction between models as surrogate systems and models as substitute systems (Mäki 2009b: 36). The important dividing line is over how much the model is required to represent the world beyond itself: that is, the extent to which economists' mathematical market models must capture the essence of everyday economic existence. Samuelson (1983 [1947]: 4) insisted that the goal should be the generation of "operationally meaningful theorems", because in their absence he doubted that he was producing successful epistemic tools. He recognized that everyday life was too complex in its layering to allow for all experiences to be represented in a single hypothetical mathematical model. Some restrictions relative to external reality were always necessary. Nonetheless, he believed that his models should be subjected to validation tests using up-to-date empirical techniques. The real world, in other words, had to be present within the model, if only in suitably idealized form. Arrow and Debreu, by contrast, made no such demand of the activity of model building. Their models at most could be said to stand in for the real world, but in being placed strictly in parallel to it they merely confirmed its absence. There is no direct counterpart to Samuelson's operationally meaningful theorems in Arrow and Debreu's approach to theoretical construction through proof-making, because there is no similar commitment to validation tests. Samuelson's models look more like what today are called surrogate systems than Arrow and Debreu's obviously substitute systems, but even they are not wholly surrogates.

Surrogate systems operate on the basis of an isolationist logic (Mäki 2001b: 373). The objective is to reveal the essence of a causal mechanism in the world to which the model speaks by allowing its counterpart in the model world to act alone, suitably cleansed of disturbing factors. A doppelganger effect is sought. If the mechanism of interest in the model world can be specified sufficiently closely to how the same mechanism is known to work in the real world, then the same impressions it leaves in the model world can be inferred to the real world. Those outcomes are only observed in the model, but if the structure of the model is itself constituted through observation of the external reality, then it is as if the two sets of observations act in unison. However, the qualifier "as if" is important here. The model can provide insights into the world that lies beyond the boundaries of its construction, but we know that it is not a direct aspect of that world. The process of isolation strips away much of the context in which the target mechanism is embedded, and so what we see in the model is not that mechanism as it is in the world, but at most a pared back version of it. An important difference thus emerges between the essence of the mechanism rendered fit for the modelling activity and the complex social reality out of which the mechanism actually arises. The doppelganger effect can therefore be misleading, because however much empirical content is poured into the features of the model, it is still a representation of the world that appears within the model and not an aspect of the world itself.

This is because the isolationist logic underpinning surrogate systems responds to a method of idealization. This notion received an early explanation by the philosopher Leszek Nowak (1980). He argues that a model formalizes a theoretical conjecture by providing an apparatus in which that conjecture can be presented in its most elemental form. The greater the idealization of the features of the explanatory model, the more elemental the form of the explanation. Aspects of the model's structure will be set to their limiting values, which will enable them to be excluded from the theoretical claim that the model is designed to demonstrate. This process can be reversed at a future date through incremental de-idealizations (McMullin 1985: 261). The model is thus refined in a more realistic manner to test whether it continues to approximate known trends in the target (Cassini 2021: 90).

However, Samuelson was no more sympathetic than Arrow and Debreu to the process of de-idealization. He asked himself how many of the most cherished theoretical principles associated with economists' market models would be disavowed were they to be shown to be empirical non-starters under less elemental versions of those models, and he could not bring himself to conclude that many of them would be voluntarily discarded (Samuelson 1983 [1947]: 117). This attitude immediately restricts how far isolations can be used to produce genuinely surrogate systems. Paying methodological lip service to operationally meaningful theorems is not at all the same as having demonstrated the practical steps that need to be taken to advance to a surrogate approach to model making.

Nonetheless, the broader point remains that it is possible to bring an isolation into closer contact with observable features of external reality through a strategy of de-idealization. Economists are thereby confronted with a choice. How much of the structure of their model will they permit to display limiting values that reduce the hypothesized causal mechanism to its idealized form? Philosophers of science tend to treat the process of de-idealization as generically a good thing. Samuelson is hardly alone amongst the economists, though, in suggesting that de-idealization can be sidestepped as a strategy if it is likely to be at the cost of more pristine versions of the market model. The pioneers of today's dominant class of macroeconomic models, the Nobel laureates Finn Kydland and Edward Prescott (1996: 73–4), have made the rather bold claim that conventional econometric estimations of model parameters are redundant for their purposes. They see no reason to try to give presence to known aspects of the external reality within their models, and they are content to give model variables unrealistic numerical characteristics if this delivers crisper theoretical implications.

The philosophy of science literature, by contrast, places great store on the representational attributes of what it calls Galilean idealizations. These are designed so that the isolated mechanism will replicate how we believe its real-world counterpart would behave were it possible to observe its operation in a completely unadulterated form. A purer version of the real world than we should ever expect to experience in practice is thus made to appear in the model (Raisis 1999: 158). For instance, the dynamics of free fall in Galileo's famous study of gravitational impulses were initially the same as those for a body released in the real world with zero velocity in either the vertical or horizontal plane. After the moment of release, though, the resistance of air acting contrary to the gravitational force on a falling body is deliberately ignored so that free fall in the model is actually freer fall than in reality (Dobson 2014:

282). Physical laws based on Galilean idealizations will typically share this property of being somewhat false with respect to real experience but overwhelmingly true enough to provide reliable explanation in a genuinely substantive manner (Niiniluoto 2014: 75). Something as close as Galileo was ever likely to get to the world beyond the model was thus carried experimentally into his model.

However, despite some use of experimental techniques today (see Conclusion), it is extremely rare for economists to create experiments along similar lines to their counterparts in physics. Jevons did try to ground his market models in this way in the 1870s, but nobody else featured in the book's historical chapters ever gave the possibility a second thought. The more that the mathematical mode of reasoning displaced the verbal, the more that the process of Galilean idealization was subordinated to the encroachment of mathematical postulation. Samuelson never departed from pen-and-paper mathematical exercises to get his hands dirty with deep-seated empirical content. His high-level commentary might have suggested otherwise – "All sciences have the common task of describing and summarizing empirical reality. Economics is no exception" (Samuelson 1952: 61) – but the collection and subsequent use of data was clearly not something he was willing to allocate scarce research time to. Whatever surrogate systems might mean in the case of economics, the isolations on which they are based are not produced in a manner analogous to Galileo's famous experiments to explain the motion of bodies experiencing constant acceleration.

We are therefore faced with something of a dilemma. Philosophers frequently insist that Galilean idealization is the most reputable form of scientific modelling, but can we really conclude, hand on heart, that this is how economists proceed? An anecdote recounted by Nobel laureate Robert Lucas is more than enough to suggest that the answer is almost certainly going to be "no". In one of the many interviews he conducted following the award of his prize, he recalled the moment when his Chicago colleague and fellow Nobelist Edward Prescott sent him a piece of paper, containing nothing other than an elaborate equation and a single sentence saying he was sure he now understood how to model the underlying logic of the labour market. "The normal response to such a note, I suppose, would have been to go upstairs to Ed's office and ask for some kind of explanation. But theoretical economists are not normal", confessed Lucas (cited in Breit & Hirsch 2004: 293), "and we do not ask for words that 'explain' what equations mean. We ask for equations that explain what words mean. Ed had provided an equation that claimed to explain how labor markets work. It was my job to understand it and to decide whether I agreed with this claim."

As Nancy Cartwright (1995: 279), a firm advocate of Galilean idealization, has argued, this technique works best in the search for universal laws of behaviour, but Lucas's story shows that theoretical economists will not usually contemplate such activities, because their attention is concentrated on their self-made model worlds. Even if this were not so, the economic world in any case throws up significantly fewer universal laws than the physical world. Furthermore, even when a universal law might be in sight in economics, it is unlikely to reduce to a simple causal relationship that might be studied separately from a broader set of interdependent causes (Alexandrova 2006: 173). The choice would therefore seem to be between, on the one hand, focusing the philosophical reflections on a standard that may forever be out of economists' reach and, on the other hand, focusing on what they are much more likely to be doing even if this undermines the clarity of the philosophical outlook.

ECONOMIC MODELS AND THEORETICAL ASSUMPTIONS

The dominant motif of the philosophy of science literature on explanation has been faith in the possibility of direct resemblance between model and target. However, some philosophers of science have proposed alternative standards. As long ago as the 1960s, Mary Hesse (1966) argued that the activity of modelling was most suited to the stimulation of analogical reasoning. William Wimsatt (1987) extended Hesse's argument to suggest that there was no requirement for analogical reasoning to shun generative assumptions that are self-evidently false with respect to the world beyond the model. Ronald Giere (1988) pushed the same position further still in saying that models could possess explanatory potential even if they were entirely self-contained and solely suppositional systems. Hesse, Wimsatt and Giere were not writing about economics specifically, but economists have clearly taken heart from the idea that reasoning from a model does not necessarily mean arguing through substantive likeness (Galbács 2020: 292). They have seemed eager to accept the invitation to be working analogically (Lucas 1981: 272). Yet still the philosophical debate about the nature of economic modelling has been set in a context that has emphasized the objective of representation, even when economists make no claims to have internalized such a standard.

Two-place analyses work on the presumption that there is an essential commonality operating between model and target (Knuuttila 2011: 265). Within this conventional framework, successive layers of reality are removed by assumption so that something more manageable emerges for the purposes of modelling. However, as Sugden (2009a: 17) notes, economists are just as likely to add things to their models that are not observed in the world as they are to remove things that do feature there. Mauricio Suárez (2010: 96) consequently argues that it is necessary to separate the representational force of the model from the inferences it licences. Economists appear to be comfortable operating in the philosophical tradition handed down from Hesse, Wimsatt and Giere, whereby they make inferences to the world beyond the model even if that world is difficult to detect within the model being used for inferential purposes. This says nothing as yet about *how* such capacities are grounded, but it does begin to move the philosophical discussion closer to what economists actually do.

According to Michael Weisberg (2007: 209), the debate about how to refine a model following its initial construction is finalized prior to consideration being given to its ability to explain anything about the real world. In other words, the crucial tests of economic models relate to their internal validity, or how persuasive they are as standalone entities to a self-selecting audience, and any attempt at demonstrating their inferential capacities takes place in a further, detached step. The primary audience for an economic model is situated far away from the philosophers who have decided that representation should be the benchmark for successful modelling. It is the modeller's fellow economists. It is less important in this regard for the model to display known real-world characteristics than for it to reinforce commonly agreed theoretical principles (Bailer Jones 2003: 66). The model will be founded on theoretical assumptions that align with the profession's internal view of what is required to be doing "proper" economics. It will then be analysed in line with how those theoretical assumptions might be further refined. If inferential capacities are to be claimed, then it will be on the basis of how well the model contributes to the development of economic theory, not how well it represents the world that people experience beyond the confines of purely academic debates.

This is the essence of the all-important difference between surrogate systems and substitute systems. The fact that hypothetical mathematical modelling in economics takes place largely as the creation of substitute systems therefore constitutes a radical break with the conventional idea of science: namely, that it builds nomological machines to help us understand what happens when certain causes operate in specific ways (Cartwright 2002: 137). Substitute systems do not provide an explanation of anything *per se*, at least not anything beyond their own internal features (Sugden 2009a: 4). Lucas (1988: 5) is entirely upfront about this limitation, describing the objects created in substitute modelling practices as artificial worlds: "This is what I mean by the 'mechanics' of economic development – the construction of a mechanical, artificial world, populated by the interacting robots that economics typically studies". Yet a model of this nature, if successful, can still enforce meaning of its own (Grüne-Yanoff 2009: 85). This is not meaning about the world in any direct sense, only meaning about how we are prepared to view the world from a certain theoretical perspective. Hypothetical mathematical models in economics are hence perhaps best seen as specially designed artefacts created for the purpose of facilitating claims consistent with a particular worldview (Ireland 2003: 1624). Model-permitted claims typically proceed by way of inference from assumptions to conclusions, both of which conform to professionally embedded norms (Kuorikoski & Lehtinen 2009: 120). It is usual from the standpoint of philosophy to think of substitute systems as free-floating tools of enquiry, because their structure exists independently of the structure of the actual economy (Morrison 1999: 64). Yet they are the very antithesis of free-floating from the standpoint of economics, because they are firmly tethered to the assumptions that underpin mathematical market models, assumptions that will have been imparted to economists from their earliest undergraduate courses.

To make sense of the content of hypothetical mathematical modelling in economics, then, we first have to understand the limits that theoretical assumptions place on the types of world that might appear in economic models. This much has been well known for some time. At the start of the period in which economics began to become a modelling science, methodologists of the subject field were making the case for falsificationist approaches that leant heavily on Popperian philosophy (Souter 1933a, 1933b; Hutchison 1938). However, Karl Popper himself harboured lifelong scepticism that economics could travel in this direction. The methodologists' conversion to falsificationism urged a programme of research in which theoretically grounded propositions would be subjected to empirical testing to see how long they could survive. It is still a cause of regret to some that no tangible progress was made in rounding out this vision (Blaug 1992: 241). There has certainly been a reaction against the other-worldly appearance of proof-making by logical assertion amidst a recent embrace of applied economic analysis (Backhouse & Cherrier 2017: 5). Yet still this has left the theoretical foundations of mathematical market models very firmly in place. Applied economists have demonstrated the often-convoluted practices of price formation operating in individual markets and have cast doubt on the process through which individual consumption and produc-tion decisions are conventionally scaled up to market demand and market supply schedules. However, these studies have not taken on the character of a new Galilean idealization that establishes alternative theoretical foundations through experimentation.

Popper (1976: 102) believed that economists' professional deference to a research agenda of competitive, price-taking, equilibrium market systems provided a means only to enact "situational analysis". The most that could be expected under these conditions was for economic theory to define the features of the fictitious world in which its starting assumptions hold, before then explaining the pattern of behaviour that one would expect to see in that world. The assumptions define the nature of the world under investigation, and it is this world that subsequently imprints itself on the decisions of those who act within it (Hay 2004: 48), not the other way around (Gelfert 2011: 283–4). The agential context cannot itself be subjected to Popperian principles of falsification, because it must be treated as a given of the background theo-retical system (see Introduction). Popper's description of situational analysis therefore looks a lot like the much more recent attempt to describe the typical economic model as a substitute system. The theoretical assumptions on which economists' situational analysis of mathematical market models are founded stretch deeply into their professional psyche, but very few would accept the claim that this is how the world actually works (Rosenberg 1995: 354). Those assumptions therefore do a lot of heavy lifting in determining the character of the model world.

Hypothetical mathematical models contain aspects of what economists are already conditioned to know (Winther 2006: 709) and allow them to make claims that they already suppose to be true (Knuuttila 2009: 76). This knowledge, though, comes from their immersion in economic theory rather than from their immersion in surrounding concrete realities. Consequently, model results are likely to be severely overconstrained: the conclusions the model ultimately reveals are already heavily implied by the structure of its assumptive base (Godfrey-Smith 2006: 734). There is an incestuous loop in operation here between founding intuitions of what is likely to make the model meaningful in the first place and what the model is allowed to confirm after the fact as actual knowledge. It is not permitted to produce theory-contradicting results, because otherwise it would fail its artefactual purpose. The assumptions therefore provide content for the model *and* all conceivable meaning for the model's results once that content has been created.

This, though, does not yet tell us about the nature of the assumptions involved in the process of economic modelling. One potentially important starting point is Alan Musgrave's (1981: 378) distinction between negligibility assumptions and what, under Uskali Mäki's (2000: 324) reworking, later became known as applicability assumptions. In Anna Alexandrova's (2006: 180) alternative terminology, they can be grouped together as situation-defining assumptions, deliberate restrictions intended to establish the nature of the world within the model. They are what assign particular behavioural capacities to the people who populate the model world, while also limiting their worldview only to that which enables them to obey a pure economic logic. Situation-definers ensure that the model world has a structure of type X and not of type Y, as well as that all behaviour occurs in a manner consistent with X and not Y.

Musgrave's situation-defining negligibility assumptions allow for an effect that is known to apply in the real world to drop out of the model, because that effect is considered unimportant for the purpose for which the theory is to be used. Note that this is not unimportant to how the world might be experienced by many people, only unimportant to the higher explanatory purpose to which the theory is aligned. As an example, psychological and sociological research has long shown that the relationship between price and demand is not a smooth inverse relationship when consumption decisions perform an act of cultural signalling. Such behavioural effects are assumed

away, though, when the model is required to illustrate theoretical propositions related to equilibrium-oriented, market-clearing dynamics. Meanwhile,
Mäki's situation-defining applicability assumptions allow for effects to be either
removed from or added to the model if this permits the theory to be applied
more rigorously to the hypothetical world under construction. For instance,
almost all modern microfounded macroeconomic models assign rational
expectations to the entire population of economic agents, because this is what
is required if theories of information asymmetry between policy-makers and
the public are to produce the sort of business cycle dynamics we are familiar
with in real life. Such behavioural effects are ascribed to the world within the
model so that the model can respond, once again, to theoretical propositions
couched in equilibrium-oriented, market-clearing terms. They are considered
indispensable to the story the theory is designed to tell.

Frank Hindriks (2006: 411–14) adds tractability assumptions to the
Musgrave-Mäki typology, which are described by Alexandrova (2006: 180)
as derivation-facilitators. In their absence, the model might yield no obvious results, because it is unlikely to be solvable. Their primary contribution is to ensure that the model itself can be closed, so that it embodies a
mathematically-determinate system of equations. Tractability assumptions
have been required at various points in the history of the physical sciences
because more complex models could only have been founded on mathematical techniques that had yet to be developed. For instance, Newton knew full
well that the solar system contained multiple planets and that each was likely
to affect all others' orbits around the sun, but he assumed away all but one
when constructing models of planetary motion, because the mathematics
of interdependent orbits was too complicated for the state of contemporary
knowledge (Scott 2007: 19). Hypothetical mathematical models must permit
a mathematical solution, and to enable the system of equations to deliver a
single comprehensible answer considerable licence is often taken with the
realisticness of the assumptive base (Carrillo & Knuuttila 2021: 53). This is
not to allow the model to speak more clearly to the theoretical propositions
whose illustration is its sole purpose – negligibility and applicability assumptions are relied upon to do that – but because the model's insights cannot
be translated into a recognisably economic claim if it is unable to first produce a mathematical solution (Hindriks 2006: 411). We are therefore likely
to make progress towards identifying the essence of mathematical market

models by focusing on the epistemic work undertaken by their tractability assumptions.

However, this does not mean that the issue has suddenly become straight-forward. It is far from clear that a strict dividing line can be preserved between negligibility and applicability assumptions on the one hand and tractability assumptions on the other. After all, assumptions do not come with an announcement about the precise job they are doing in the construction of the model. They are merely presented as givens of that particular structure. Take the assumptions of homothetic preferences that deliver utility functions which are homogeneous of degree 1; the independence of preference functions; the representative individual who stands in for the whole population; a downward-sloping demand curve and an upward-sloping supply curve; homeostatic motion that eventually nullifies exogenous disturbances to the system; and the existence, uniqueness and stability of the condition of equilibrium. All of these might be seen as assumptions that allow the theoretical propositions illustrated by market models to take pristine forms relieved of the presence of alternative causal mechanisms. Yet they could just as well be seen as assumptions that allow the model to have a mathematical solution. So, which are they, situation-definers or derivation-facilitators?

Despite these identification problems, it is worthwhile persisting for heuristic purposes with the distinction between situation-defining and derivation-facilitating assumptions. As hypothetical mathematical modelling has evolved in economics, the general trend has been for more emphasis to be placed on the mathematical than on the economic requirements of the model (see Chapters 6 and 7). As a result, the relative importance of the work being done by individual assumptions has shifted from defining the situation of the world within the model to facilitating the derivation of the model's solution. By investigating further the significance of this latter class of assumptions, we can see how far economic theory has come in its modelling endeavours from exhibiting a representational logic to exhibiting an artefactual logic. Under artefactual logic, however, the explanatory unification on which economics imperialists pride themselves stops at derivational unification (see Chapter 1).

DERIVATIONAL ROBUSTNESS IN ECONOMIC MODELLING

Robustness stands alongside rigour, precision and generality as an economics imperialist's go-to justification for the extra credibility that should be given to their approach. It is a virtue term, where the more robust an explanation the more persuasive the claims that can be made in its name. For a supposedly crucial distinguishing feature, though, it remains remarkably poorly defined in the few explicit mission statements delivered by card-carrying economics imperialists (see Chapter 1).

Following Edward Leamer (1983: 38), a general sense prevails that robustness is facilitated by removing results that are overly sensitive to a model's particular specification. If a class of models produces results that follow the same clear pattern under alternative specifications, then that pattern can be considered robust. But what is it that counts as the factor of concern that should ideally be insensitive to alternative specifications? Is it the content of the numbers determining the shape and the size of the motion that sets in train economic outcomes in the model world? Or is it the broader structure that provides the model world with the appearance of economic motion in the first place? There seem to be two distinct routes here through which robustness can be approached: inferential robustness in the former instance, derivational robustness in the latter.

A discussion around competing claims to robustness can quickly become muddled if it does not make it clear which of inferential or derivational robustness is being referenced. However, much of the methodological literature proceeds as if inferential robustness is all there is (Woodward 2006: 219). As economists are not normally in the business of representational inference, does derivational robustness fare any better as a standard that most would accept as their own? It focuses on internal comparisons of the precise specification of different models of the same overall type. Derivational robustness is about comparing different expressions of the world within the model. The generic definition of robustness is likely still to hold in these circumstances: namely, the attempt to ascertain how stable the result is in the presence of different assumptions. Yet in this instance the notion of "result" has changed relative to its location in the description of inferential robustness. There, the result belongs to the world beyond the model once the process of inference from the model has been allowed to run its course. Here, the result belongs to the world within the model in the absence of the process of inference. Derivational robustness thus proceeds through the creation of multiple models that each capture the same sense of how a hypothetical economic system responds to exogenous pressures even though that system is specified differently in each instance.

This is a relatively new approach to the question of the epistemic value of hypothetical mathematical modelling in economics. Interestingly, the practice was well established long before any justification for it began to take shape. The philosophical literature has focused on filling in the backstory to make sense of what economists were already used to doing. The chief proponents of the epistemic value of derivational robustness are Jaakko Kuorikoski, Aki Lehtinen and Caterina Marchionni (2010, 2012). It features as part of their theory of modelling as extended cognition, whereby "merely looking at models may justifiably change our beliefs about the world" (Kuorikoski & Lehtinen 2009: 130). Sugden (2000: 24) had earlier argued that economists could make plausible inferences to the real world on the basis of hypothetical mathematical models containing purely imaginary populations, on the sole condition that their model worlds were credible. By this he did not mean that they had to mimic the conditions that were visible in any known world, only that it did not stretch the imagination beyond breaking point to envisage circumstances in which those conditions might apply (Sugden 2009a: 17).

Sugden (2011: 718) has characterized most hypothetical mathematical models in economics as explanations in search of a corroborating observation. The fact that explanation might precede observation acts as a warning to expect slippage between model-world and real-world relationships. But still he stresses the positive epistemic function of modelling in such a way (see also Boumans 1997: 69). Kuorikoski, Lehtinen and Marchionni are equally adamant that much of real-world significance can be learnt through the modelling process, but if Sugden requires significantly less of economic models than the standard of Galilean idealization, they require less still. For a start, their conception of model robustness entirely bypasses the need for successful inference. From their perspective, Sugden's credible worlds approach comes too close to heading back down the surrogate system cul-de-sac that is a hangover of the subject field's historical physics envy (see also Knuuttila 2009: 65). Once released from such pressures, they say, models can stand on their own as tools for learning, however incredible they appear to those who still believe that the real world should be detectable at least to some degree in the model world (Kuorikoski & Marchionni 2014: 322). At the same time, though, this epistemic function should not be confused for explanation *per se*, because at most the postulated elements of mathematical models can facilitate formal understanding, or "pseudoexplanation" as Kuorikoski (2021: 190) puts it.

Most economic models work by deriving results X_1, X_2, X_3 ... X_n consistent with what economists' trained intuitions tell them they can see in the world and with what their mathematical capabilities allow them to make visible. Certainly, this is how economics imperialism works. Economics imperialists model all situations of social interaction as a market problem. Each ensuing mathematical iteration of the basic market model will differ from one another, even if not greatly so, because they will contain subtly different starting assumptions that require slightly different mathematical treatments. Model development occurs over time through the modification of the assumptive base (Hands 2016: 34). But will the derived results X_1, X_2, X_3, etc. still be X_1, X_2, X_3, etc. when assumption A_1 becomes A_2 or assumption B_1 becomes B_2? If the derived results X_1, X_2, X_3 ... X_n are still either very clearly the same or sufficiently close to being so that it would be splitting hairs to emphasize their difference, then these results represent a robust derivation. However, if X_1, X_2, X_3 ... X_n become X'_1, X'_2, X'_3, ... X'_n as soon as the assumptions change, then the derived results cannot be said to be robust (Woodward 2006: 231). For this definition of robustness to hold, though, we would have to be operating very much on the metamathematical territory of defined mathematical functions.

Derivational robustness involves subjecting to systematic enquiry the effect that particular modelling assumptions have on modelling results, not the effect that those assumptions have on what we might be able to explain of the world beyond (Kuorikoski, Lehtinen & Marchionni 2010: 541). It allows for triangulation via repeated mathematical refinement of the assumptive base, rather than moving through different levels of abstraction in the search, ultimately, for the concrete instantiation of results (Weisberg 2013: 168). Proponents of this technique celebrate its achievement in being able to show which element of the mathematical structure of the model bears greatest responsibility for the precise character of the results (Weisberg 2006: 737). Sceptics, meanwhile, remain concerned that it is much more likely that situation-defining assumptions will be subjected to robustness checks than derivation-facilitating assumptions, and that it is the unchecked assumptions that render the model mathematically tractable (Cartwright 2007: 225–6).

What, then, are economic models models *of*? They must be more than their mathematical content, because it can only help identify a mathematical solution to a mathematical problem. Even if the dynamism of the model is concentrated solely in the manipulation of its mathematical characteristics, there must be a feature that turns the mathematical modelling process into the more specific exercise of mathematical economics. Marcel Boumans (1999: 90) suggests that this extra step is a process of "moulding", through which the mathematics allows for different theoretical elements to be integrated into a single model. However, this serves merely to push the problem down the road, because it is still the *mathematics* doing the work for what is being presented as an *economic* model. The translation of theoretical propositions into something that is mathematically mouldable must be translated again, this time into something that is economically meaningful. Without this second step, whatever the economic model is a model of, it is not something that has any obvious economic meaning relevant to everyday experience.

Paul Krugman (2009: 4) has drawn the important distinction in this regard between the scaffolding and the narrative of a hypothetical mathematical model. The scaffolding is the intellectual edifice of a system of equations that delivers a unique solution, while the narrative brings the system of equations to life as something more than itself to say why that solution holds relevance for the way real-world economies are organized. It is the storyline accompanying the mathematical structure that elevates the model beyond being a wholly self-referential account of the structure's components. This is what turns it into a "model of something" rather than just the working through of the innate logic of the hypothetical artefact (Grüne-Yanoff & Schweinzer 2008: 133). Of course, the activating story is no less an artefact than the mathematical structure is. It is also an invention, but it will feel more familiar as it reflects the theoretical propositions that trigger the search for the most suitable mathematical outlet (McCloskey 1983: 505). It will be presented in ordinary language and, as such, it will make fewer demands on the technical skills of the audience. The closer the storyline comes to presenting something that does not really need to be said because it reflects what other economists would already accept to be true, the more instinctively plausible it will sound. The economic modeller therefore aims for believability in two distinct ways.

It is important for the mathematical structure to pass the judgement of other modellers; it must look right as a model. But it is also important for the accompanying storyline to be accepted; it must feel right as a specifically economic model.

The storyline has two functions (Morgan 2001: 366, 380). It defines the question on which the model is asked to provide insight, and it also injects essential elements of economic sense-making into systems of equations. Market models define every question as an equilibrium problem: how will rational agents in any given circumstances maximize their objective functions? At the same time, the mathematical solution to the equilibrium problem cannot be just a mathematical solution. It shows that there is reason to believe in the possible existence of an equilibrium point, and the storyline is then called upon to ensure that there is an economic explanation of what that equilibrium might be taken to mean. Note, though, how the same assumptions feature in both aspects of the model, the scaffolding and the narrative. For instance, every behavioural trigger that does not conform to a strict interpretation of purely instrumental rationality is eliminated from consideration. However complex the market model, it requires this restriction if the model is to produce a unique solution and prove that the structure is mathematically tractable. It is a derivation-facilitating assumption when used this way. That same assumption is also what allows the storyline to explain in economic terms why the internal motion of the world within the model is towards equilibrium and not away from it. It will only allow economic agents to think and to act in one way, and in these moments it is a situation-defining assumption.

The storyline embedded in an economic model therefore links the model to both a set of theoretical propositions and a particular system of equations. It consequently seems to belong only to the world within the model and has nothing to say about the world beyond. However, there is one additional complication still to review. The defence of the epistemic function of derivational robustness in economics typically follows from asserting that the model storyline itself carries empirical content (Hands 2016: 45). This might not be tightly defined empirical content (nobody is saying that actual people respond to their behavioural context in exactly the way the market model requires), so much as that it conforms to stylized facts (economists are often happy to conclude that this is sufficiently close to actual economic behaviour to stand in for it). Some essential version of the world beyond the model thus becomes visible again within the model world, even if it is not the version that anyone encounters in their everyday lives. It gains a presence there through the stylized facts that economists are willing to accept as an element of the narrative counterpart to their mathematical structures. The storyline continues to be based on fictions, because it is not saying that this is exactly how the world beyond the model works, but these are fictions suffused with some kind of realism, or so the defence goes (Davies 2007: 29). In the same way that novels depict purely hypothetical scenarios but render them believable if they appear familiar to actual circumstances we can imagine ourselves experiencing, so it is said to be with economic models (Frigg 2010: 257; Pignol 2023: 17). From this perspective, nothing within the model must be literally true for the storyline to be true enough to embed the image of the real world within the model world.

We are thus drawn back again to the issue of inference. As a matter of routine, economists use the language of inference when discussing the significance of their models, and in its absence economics imperialists would struggle to make the case for their approach. Take that heightened inferential capacity away and what does it do to the demand for hypothetical market models to be introduced to subject matter throughout the whole of the social sciences, if not pull the rug decisively from beneath it? Economics imperialism is likely to continue to be founded on assertions of its greater inferential potency, because without such assertions it would be very difficult to shrug off accusations of the emperor's new clothes. However, herein lies a problem. As a rule, economists do not build their various iterations of market models with inference in mind. Their commitment to the construction of substitute systems as self-standing artefacts bypasses altogether the modelling realm in which philosophers of science tell us inference operates. Economists tend to use the language of inference but without using the relevant model characteristics to allow inferential capacity to be built into the explanatory schema. Economics imperialism would therefore seem to be based on developing substitute market systems but then treating them as if they were surrogate market systems, with the formal mathematical structure acting as a distraction device designed to cover up such sleight-of-hand. This move appears to be philosophically untenable, but believing in it is crucial to the ongoing success of economics imperialists' colonizing activities.

CONCLUSION

Mathematical market models are offered as a universal means of unlocking knowledge across the whole of the social sciences, because they allow for extra clarity, precision and rigour in explanation. However, there have been two distinct trends that have shaped the way in which the mathematization of market models has changed over the last 150 years. In the previous chapter, I studied the effects of argument by mathematical postulation on both the structure of market models and what those models can be called upon to say about the world beyond themselves. In this chapter, I have undertaken exactly the same exercise but this time in relation to argument by hypothetical mathematical modelling. Each has shown the significance of the mathematical manipulation of systems of equations to how economists think and reason with their theoretical propositions. But each has also shown that mathematical treatment of economic activity repositions that activity as something which privileges an abstract internal essence over context-specific behavioural choices. Actual economic behaviour consequently seems to have been placed at an ever greater distance from the construction of the market models that supposedly explain that behaviour. Yet economics imperialists demand not only that we overlook the possibility that the development of an overtly uneconomic economics is problematic for explaining economic behaviour, but also that we should accept that it explains non-economic behaviour better than theories that were designed for that specific purpose.

Such tensions are seen perhaps most obviously in this chapter in the rather tortured debate in the philosophy of science about the epistemic value of hypothetical mathematical modelling in economics. Philosophers have various explanations of what might be learnt from scientific models in general, but every time those arguments are transposed to economics they appear to fall short. The underlying issue is how to speak with authority about the world beyond the model if that world is not obviously represented within the model. Economists typically sidestep the representational logic that acts as the usual benchmark of good explanations in the philosophy of science. Yet it is only under the influence of such a logic that the real world can clearly be said to hold a palpable presence in the model. Otherwise, it is difficult to know exactly what hypothetical mathematical market models are models *of*. They function perfectly well for working through the implications of assuming into existence as a thought experiment an abstract market-bound world, but how much more can they do than that? Whatever bridge exists between model-world and real-world dynamics is typically activated by the model's accompanying storyline. Yet this is only likely to provide the mathematical structure with an economic interpretation consistent with the subject field's accepted theoretical starting points.

The issue of explanatory unification thus re-emerges (see Chapter 1). What are we looking at if very similar mathematical models are made economically interpretable through the use of one narrative and legally, socially, culturally or politically interpretable through the use of other narratives? This is what happens under the influence of economics imperialism through the imposition of ostensibly the same market model across multiple areas within the social sciences. It is only the accompanying narrative that obviously changes, as the explanation in each case is contained within generically the same mathematical solution to a system of equations. In the hands of economics imperialists this can be an explanation of whatever they wish to apply it to, but only if the real phenomenon under discussion is first transposed into a market frame of reference and subjected to market logic. The old maxim is relevant here that to a person armed only with a hammer everything begins to look like a nail. To a person armed only with a market model everything must first be made to look as if it is a market entity. The real phenomenon under investigation is required to change form so that it becomes amenable to an explanation that is already predetermined. The nature of the explanation is, in effect, set in stone (see Introduction). It is not too difficult to see in this regard why Sugden (2011: 735) suggests that many applications of mathematical market models are explanations in search of an observation.

The explanatory unification enacted by the transgressive logic of applying market models to non-market entities thus falls a long way short of ontological unification. It would therefore seem to be an example of what Mäki (2009a: 373) calls "bad" economics imperialism. The narratives which bring mathematical objects to life with legal, social, cultural or political significance appear to be talking about real phenomena. Yet they do so many stages removed from how we might expect to encounter those phenomena in everyday life. The narrative content of boundary-hopping market models presents an account of what non-market aspects of reality would look like were they to adopt a form unknown outside hypothetical market relations and respond to nothing other than hypothetical market price signals. This replacement of their actual type by something they are not shows how far we have travelled from surrogate models that attempt directly to represent real phenomena. Such a move facilitates a solution to the system of equations on which the mathematical object is constructed, but the solution concepts refer only to life as it is required to be lived by the assumptions of the market world. We are consequently faced with explanatory conditions of derivational unification at best.

The philosophically oriented chapters are now at an end. They have provided numerous layers of commentary on the difficulties involved in developing a philosophical defence of economists' use of mathematical market models and, by extension, a philosophical defence of economics imperialism. The remaining chapters are written in a more obviously historical register, to see whether more plausible justifications of that practice lie there. They focus specifically on those moments in the prehistory of economics imperialism in which the basic outline of modern mathematical market models began to take shape, looking for the concessions that had to be made once it became obvious that hypothetical mathematical models were no place for real economic agents (see Chapter 1). Is there evidence that the pioneers stumbled across the same issues that have latterly proved so troublesome for philosophers of science? Did they ask themselves how epistemic value can be extracted from mathematical market models that increasingly operated in parallel to the experience of real-life markets in the transition from surrogate to substitute models? What limits were they conscious of introducing into their modelling endeavours, and what do those limits tell us about the contemporary practice of economics imperialism?

CHAPTER 4

Marginalism as proto-imperialist move 1: Stanley Jevons and the search for enumerated principles of economic behaviour

INTRODUCTION

Most proponents of economics imperialism show scant regard for trying to make their case historically. Declarative statements of the "it has long been known" variety serve their purpose much more effectively. Thus, if it has long been known that the most tractable models of human behaviour are based on maximizing principles, if it has long been known that the economic content of maximization revolves around individuals following their own rational self-interest, if it has long been known that mathematical expression brings extra precision to the economic content, and if it has long been known that good social science requires such precision, then this is often enough to negate the case against the imperialists. In a single step, it seems, the debate shifts from the nature of explanation in general to the assertion of explanatory unification through the use of mathematical market models. No concern is given for what type of unification is entailed, nor yet for what particular market models assume about the external reality that is actually unknown there. A rush to judgement ensures that the market world is always a credible world in Robert Sugden's (2000: 24, 2009a: 17) sense, thus licencing an inferential capacity that far exceeds what subject specialists can claim of their own work.

It appears to make no difference that these four "it has long been known" claims came to economics at different times, were hotly contested in methodological terms in their own day and have remained significant points of contestation. What counts is that they are widely considered to be true, beliefs that are embedded when learning how to think like an economist. Economic models are often constructed to test theoretical claims that are already assumed to hold (Ireland 2003: 1624; Winther 2006: 709; Knuuttila 2009: 76; see Chapter 3). This leaves unexplained, though, how such assumptions formed, especially in the face of historiographical evidence that the authors with whom they are most usually associated all seemed to harbour reservations about the advances with which they are now credited.

I identify the first phase of the prehistory of economics imperialism – proto-imperialist move number one – with the marginalism of the 1870s. Along with almost all other important instances in intellectual history, this was a moment of codification rather than of pure invention, as there were clear pre-emptions for the previous 40 years. Even when it was given its first systematic treatment by the pioneers – Stanley Jevons (2013 [1871/1879]), Carl Menger (1981 [1871]) and Léon Walras (1984/1954 [1874]) – there were innumerable cracks in the edifice of scientific unification. The economists of the 1870s did not speak with one voice, even if as we zoom out from the finer details of their arguments the outline of a common framework can come into view. For my purposes at least, the most important instances are those where the early marginalists evidently disagreed with one another. The emphasis in the conventional account of this period is always on how united they were in their efforts to unseat an established orthodoxy. However, this will have looked different to the Anglophone Jevons, the Germanophone Menger and the Francophone Walras. They were all located in their own national tradition of economic thought and their theoretical work therefore reflected divergent influences and divergent legacies. The image of a single shift is thus extremely difficult to sustain (Jaffé 1976: 522).

Each of Jevons, Menger and Walras provided a means of systematizing economic theory through its mathematical treatment, but there were many mathematical instincts simultaneously in play. These differences were magnified by their successors and magnified again by their successors' successors. Menger is arguably the most interesting in this regard. He shunned direct mathematical expression but nonetheless wrote clearly within a style founded on mathematical logic. The second- and third-generation Austrians – Friedrich von Wieser, Eugen von Böhm-Bawerk, Ludwig von Mises and Friedrich von Hayek – replicated Menger's basic structure of argumentation. Meanwhile, Jevons was followed by the more mathematically savvy Francis Ysidro Edgeworth and Philip Wicksteed, but then by Lionel Robbins exhibiting the influence of the later generations of both the Austrian and Anglo-Irish economists. He inherited the former's wariness of direct mathematical expression but engaged with the theoretical implications of the latter's increasingly robust mathematical logic. Vilfredo Pareto worked within the inheritance handed down by Walras, but while the volume of mathematical expression visible in Pareto's work was less than in Walras's, his mathematical content was more precise. Read as a whole, each generation of marginalists removed another layer of literary style, and in its place they added forms of mathematical expression, content and logic that were deemed to be cutting-edge in their own time (Jaffé 1964: 93; Howey 1973: 21). However, the frontiers of mathematics are nothing if not a moving target, imposing a similar effect on mathematical economics. What counted as mathematical state-of-the-art for each generation of marginalists repeatedly took on the mantle of the mundane for successor generations. We are therefore looking at multiple shifts in style, tone, register, method and theory, not just one all-embracing revolution.

In any case, perhaps the only genuine revolutionary of the three founders was Jevons (Dobb 1973: 166). Walras was the most energetic in trumpeting his claims to scientific priority, but he did not have an obvious established orthodoxy to push against (Walker 2005: 423). Menger too was working within the existing European mainstream of subjective value theory (Coats 1973: 49). Only Jevons lacked the comfort of arguing on already generally agreed territory. Walras and Menger could always console themselves that if they wanted to talk about value as a subjective phenomenon centred on the mind, there was no gulf of basic meaning to first be bridged for their reasoning tools to make intuitive sense to others (Black 1973d: 100). By contrast, the text of Jevons's *Theory of Political Economy* offers significant clues to just how wide he considered the corresponding gap to be. Stephen Stigler (1982: 361) captures this perfectly when commenting on the "dramatic italicized sentences" that are dotted throughout Jevons's work. They arise every time he thought he was attempting something previously unknown and wanted to signal to his readers that they should expect to encounter what he openly described as his "sedition" (Jevons 2013 [1871/1879]: 276).

The analysis in this chapter concentrates on how Jevons attempted to enact such a break. To keep the discussion focused on the prehistory of economics imperialism, I will be especially keen to examine the exact ways in which he sought mathematical influence for his economic theory. His 1871 *Theory of Political Economy* often features direct mathematical expression as the primary means of reader engagement, making use of explicit mathematical notation in a way that represented a significant departure for the Anglo-Irish tradition. There is also plenty of evidence of mathematical content throughout Jevons's economic writings, but this was of a specific kind. It revolved around a strategy of enumeration, where he insisted that any number which was driving his economic theory was extracted wherever possible from observations of real-life economic relationships. Jevons's theoretical claims often look as though they are based on very rudimentary forms of mathematical postulation, but there also seems to be a constant urge on his part to ensure that the world beyond the model was represented in his model world, in the manner of described and not defined functions. He cannot have been expected to have used the much more modern philosophical language of resemblance between model and target, yet his work certainly hints of some such pre-emption. Number, for him, typically implied statistical enquiry of known relationships within everyday life, where experimentation played a key role in his very basic proof-making.

The chapter now proceeds in four stages to capture the specific intellectual developments on which Jevons thought he was commenting. The difference should therefore be clear with the modern-day tendency to treat the marginalist pioneers as being all of a piece. In section one, I highlight how Jevons divided the discipline into numerous discrete functions in his efforts to restore British economics into the European mainstream of subjective value theory. Abstract theoretical claims were considered the most important for the development of the science, but they were required to coexist with more practical concerns if economists were also to be socially useful. In section two, I detail how Jevons fleshed out his subjective value theory with a particular conception of economic agency. Later phases in the prehistory of economics imperialism tend to model all economic decisions on those of an abstract consumer, but Jevons's agent was always much more than that. His attempts to make it a recognisably real person meant that it was also more than a quintessentially economic decision-maker. In section three, I show how this implied a blurring of the lines demarcating economics from other social sciences. The experimental method of late nineteenth-century psychophysiology struck Jevons as being the most feasible template for developing an empirically robust subjective value theory built upon the concept of utility. Subjective value theory might well have appealed to his fascination with making economic arguments through the medium of mathematical logic, but still he wanted that theory to be grounded, wherever it could, in statistical evidence. In section four, I demonstrate that this particular concern was rendered problematic by the tenor of the nineteenth-century Anglophone free will debate. Jevons's attempted solution to the question of free will was an evident fudge, as he ended up endorsing averaging techniques across the whole of society in a manner that seemed to invalidate active volition altogether.

JEVONS AND THE POLITICS OF THE ANGLO-IRISH *METHODENSTREIT*

The Austro-German *Methodenstreit* of 1883–84, a struggle over methods but also for the very soul of economics, was fought between Carl Menger and Gustav von Schmoller. They asked how much economic theory should be required to reflect the historical conditions in which it was situated, so that it might be said to capture something essential in all economic experiences, but only insofar as those experiences were recognisable from everyday life. The lesser-known Anglo-Irish *Methodenstreit* did not follow suit in being a head-to-head set-to between an established marginalist and an established anti-marginalist position. But it was more important to the concerns of this book. The Austro-German *Methodenstreit* occurred at almost exactly the mid-point between the two defining attempts to resolve similar, albeit far less rancorous, discussions of appropriate methodology that engaged British and Irish economists. The Anglo-Irish *Methodenstreit* rumbled on from the 1860s, providing a constant backdrop to Jevons's attempts to get his new theory of value noticed amongst his most immediate peers, and it came to a far from permanent truce only in 1890, with the publication of John Neville Keynes's influential treatise, *The Scope and Method of Political Economy*. Jevons (1876: 624) had pointed the way to a potential solution in 1876, having been the first to specify a compromise whereby marginalist and historical economics might happily coexist, each enjoying its own place within the overall structure of economics.

In no sense, then, can Jevons be seen as a straightforward Menger figure within the Anglo-Irish *Methodenstreit*. Neither was his apparent adversary, Thomas Edward Cliffe Leslie, a mere Schmoller figure for Anglo-Irish historical economics. There was considerably more scope for Jevons and Leslie to meet part-way than the Austro-German warring factions. Unlike Schmoller, Leslie (2013 [1870]: 375) believed that abstract theoretical laws were an invaluable element of the economist's intellectual armoury, even if there were no aprioristic means of accessing them. Before aspects of economic behaviour could be adequately captured in theoretical form, he argued, their history first had to be understood; this was how to ensure that real economic phenomena had an obvious presence in abstract theoretical models. On this point, unlike Menger, Jevons (2013 [1871/1879]: xxxvii) deferred to "my friend Mr. Leslie", even if they had different views about exactly how much of the real world should appear in economic theory and in what way. Leslie was opposed to the projection of supposedly universal laws beyond the specific circumstances under which observation could show they applied, not to the idea that it was possible to discover some law-like relationships. Once more, Jevons agreed, with both locking horns with the contemporary British standard-bearers of Millian classical economics. "The laws which regulate the value of the supply forthcoming from the producers has been almost exhaustively developed in political economy", a young Leslie (1862: 8, emphases added) wrote at the beginning of the 30-year Anglo-Irish *Methodenstreit*, "but the *deeper laws* which regulate the demand of the consumers, and which give the love of money all its force and all its meaning, has never yet received the regular attention of any school of philosophers".

The timing of Leslie's concern is significant. "The Love of Money" was written in 1862, the same year in which Jevons presented before the British Association for the Advancement of Science an ill-fated paper, "Notice of a General Mathematical Theory of Political Economy" (Jevons 1862). Jevons's diary entries from 1860 show that he had been making rapid headway in casting off the shackles of contemporary cost-of-production theories of value, a sense of progress he felt so keenly as to be comfortable confiding to himself that he was "arriv[ing] as I suppose at a true comprehension of *Value*" (Jevons 1981: 120). The 1862 paper was his first attempt to systematize the insights of his discoveries, but if he was successful in outlining the mathematical principle of a balance between pleasure and pain the same could not be said of his understanding of the psychological attributes of the underlying behaviour. This should not be surprising. He had confessed to his brother that he was "out of his depth" when it came to the metaphysical principles that might have allowed him to turn his rudimentary mathematical structure into a genuinely explanatory framework (White 1994: 221). The unevenness between the two parts of his argument produced an underwhelming response from the rest of the British scientific community. He noted that his mathematical treatment of economic theory was "received without a word of interest or belief" amongst the conference attendees, concluding that "it is useless to go on printing works which cost great labour [and] much money [but] are scarcely noticed by any soul" (cited in Peart 2003: 2). He wanted the real world to be represented in his mathematical model, but could not see how. However, such an obvious snub from his peers was soon followed by reading Leslie's essay, which convinced him to double down on the assertion that his diagnosis of the problem had been right all along. He consequently determined that his second exploration of a subjective theory of value would be a full book-length treatment that adequately filled in the gaps in his first paper.

Classical theories of value were based on attempts to measure the exchange value of commodities, recognizing that the source of the measurement must reside in the common inputs that were embodied in the products during the process of production (Moscati 2013: 388). All commodities were therefore generically the same in how they reflected the manifestation of human labour time. What Jevons (2013 [1871/1879]: 71) proposed in his *Theory of Political Economy*, by contrast, was to treat every product as distinct, even in the context of identical goods arising from the same batch of the same production line. Each product, he argued, should be thought of in relation to the unique degree of utility it provides to the specific consumer who purchases it, and this in turn should be understood in terms of a continuous mathematical function of the quantity consumed (Trupiano 2013: 159). Jevons's developing theory of value was thus wilfully anti-materialist when compared to the dominant cost-of-production accounts (Schabas 1990: 21).

The doctrinal element of the marginalist revolution was ontologically prior to the analytical in Jevons's hands. Marginalism provided the technique for understanding value from the demand side, but this came only after overthrowing the assumption that the context in which economic theory came to life was the social struggle over surplus (Tarascio 1972: 408). Jevons objected to the idea that value theory had been conquered once it had been determined who, between the worker, capitalist and landlord, might lay claim to "own" the value embedded in a particular commodity (Clower 1998: 399). He emphasized instead the sense of struggle that individuals have with themselves when attempting to balance primordial sensations of pleasure and pain. Value was therefore a characteristic of the individual's all-round engagement with economic activities and not a characteristic of the goods that came into their possession. The pleasure arises in the moment of consumption and the pain from the act of labouring. His theory thus allows individual labourers a degree of control over their input into the production process (Jevons 2013 [1871/1879]: 178). This theoretical development must be understood within the context of an awakening middle-class conscience that was driving the demand for a new type of economics, even if those same middle-class sensibilities were typically encased in a general snobbishness towards the behaviour of working-class communities (Maas 2005a: 210). It is therefore a point worth exploring further, so that the tendency towards individualism evident in the resolution to the Anglo-Irish *Methodenstreit* can be situated against its original political backdrop.

Arnold Toynbee served unofficially as spokesperson for a middle-class reformism that seemed to touch all economists in Britain regardless of their theoretical allegiances. He believed the public had been betrayed by the heritage of unreflexive abstraction embodied in the unrelenting application in policy of universal market-based laws (Winch 2002: 5). The economics profession subsequently took up their pens in support of the respectable working classes (Sigot 2002: 266). Indeed, they did so with such passion that Toynbee (2011 [1884]: 1) felt able to conclude in the early 1880s that the "bitter argument between economists and human beings has ended in the conversion of the economists". Herbert Foxwell praised the new mathematical economics for which Jevons led the way (Winch 1973: 75), identifying within it the most plausible means of bringing humanistic feeling into economic theory (Coats 1996: 80). It was his hope that future historians of economic thought would see fit to record this as the moment at which his peers had "banished to Saturn ... the mechanical unmoral economics" (Foxwell 1887: 102). "It vexes me", Foxwell (cited in Black 1973c: 186) had written to Jevons five years previously, "to hear the authority of Political Economy always appealed to by the selfish rich". Leslie (2013 [1870]: 89, emphases in original) objected in similar terms

to the Millian heritage having turned economics into a "science *for* wealth …
[and not] … a science *of* wealth". Economic theory, he maintained, projected a
cosy landowning view of a uniform world embodied in freely drawn-up con-
tracts onto what was, by contrast, a deeply divided social structure (Moore
1995: 73).

Jevons entered enthusiastically into the chorus of disapproval. He sought to
resolve the Anglo-Irish *Methodenstreit* by acknowledging the different character traits that the economist must display: sometimes pure theorist, at other
times policy advocate. However, the content of the interventions he latterly
authorized appear to have entailed vacating the political positions to which
his younger self had subscribed. The early Jevons's default position was always
to promote the "sturdy individualism" that he had been brought up to respect
(Hutchison 1982: 367), having been socialized to "the liberal nonconformist
outlook of [his] family" (Black 1982: 425). With what eventually became the
Reform Bill of 1867 already on the cards, coupled with the more sympathetic
trade union legislation that increasingly promised to disturb the established
social hierarchies of mid-Victorian Britain, Jevons (cited in Black 1973b:
150) protested in an 1865 letter to his brother that if this is what counted
for modern democracy then he, at least, was "not a democrat". He was no
early advocate, then, of the progressive regulation of economic affairs (cited
in Black 1973a: 343).

As his thinking matured, though, it departed ever more clearly from simple
rubber-stamping of the social *status quo*. Jevons learnt how to empathize with
the living conditions of the poor through Charles Dickens's ability to capture
working-class struggles against squalor (Mosselmans 2001: 223). This provided him with a secular route to the claim for which the early Alfred Marshall
(2013 [1890/1920]: 1) required religious underpinnings: that for economic
theory to serve its ultimate purpose, it had to show how good could be done
within the world. *The Theory of Political Economy* might well have extolled
"perfect freedom of exchange … to the advantage of all" across its various
editions (Jevons 2013 [1871/1879]: 142), but Jevons's policy ideas adapted
with the times. They were always a reflection of what he believed a man like
him should have had to say about the place of society's least fortunate. The
missionary purpose that ultimately acted as the backstory to his economic
theory was of a wide-ranging benevolence, whereby his instinctive support
for *laissez faire* as an ideal was increasingly tempered in practice by pragmatic

recognition of the need to do more to alleviate poverty than simply chide the poor for their own conditions of existence (Paul 1979: 278). In an 1869 speech as president of the Manchester Statistical Society, Jevons (cited in Black 1982: 425) had complained that "the poorest classes [displayed] a contented sense of dependence on the richer classes for those ordinary requirements of life which they ought to be led to provide for themselves". Yet by 1876 that concern had been overwritten by distaste for the way in which the "rights of private property and private action are pushed so far that the general interests of the public are made of no account whatever" (Jevons 1876: 630).

The standard historical narrative of the 1870s – that of a quick and complete victory of abstract individualism at the combined hands of Jevons, Menger and Walras – therefore looks to be a poor fit with what was actually going on at the time within the English-speaking world. By the middle of the Victorian era, entry into economists' political common sense was accessed only by privileging the social over the individual (White 1996: 107). Foxwell (cited in Black 1973c: 186) wrote to Jevons shortly before the publication in 1882 of his final book, *The State in Relation to Labour*, to say: "I hope to find that you have taken up … a position from which you recognize the obligation of the individual to society". "I fancy the new book will almost exactly meet your views", Jevons (cited in Black 1973c: 187) replied. His sympathy for the English historicists is clear in his denunciation throughout this book of the defence of inequality through appeal to some universal abstract economic law. Concern for the poor in the 1870s was less a temporary intellectual infatuation amongst Anglophone economists and more a whole life situation.

In his *Political Economy* primer of 1878, Jevons (1878: 132) wrote that "what we have in political economy to look to, is not the selfish interests of any particular class of people, but the good of the whole population". The youthful Jevons who had fretted about the implications of the 1867 Reform Act held the mature position in 1882 that "all classes of society are trade unionists at heart, and differ chiefly in the boldness, ability and secrecy with which they push their respective interests" (Jevons 1882: vi). He had thereby turned himself into an advocate for the new humanistic consensus that had swept through the polite Victorian opinion that Anglophone economic theory reflected (Coats 1996: 91). The attempt to use subjective value theory to restore British economics to the European mainstream was functional to this underlying political objective. The theoretical account of everyday economic life then in vogue was not that of a scientific community of economists, he seemed to be saying, so much as that of the privileged members of Britain's class-riven society. The state of orthodox economic theory thus appears to have turned Jevons into a democrat, even though he had worried less than a generation previously that his political instincts would not (Jevons, cited in Black 1973b: 150). In the context of the preceding political capture of the objectives of economic theory, he suggested that "the general public would be happier in their minds for a little time if political economy could be shown up as imposture" (Jevons 1876: 619).

JEVONS'S ACCOUNT OF ECONOMIC AGENCY

Jevons attempted to undo the control exerted in Britain by cost-of-production theories of value, but his attack came only *after* the radical potential of the labour theory of value had already been nullified. The simpler version of that theory offered by the Ricardian Socialists – Thomas Hodgskin, William Thompson, John Francis Bray, John Gray, Charles Hall, Percy Ravenstone and Thomas Rowe Edmonds – amounted to a clear political manifesto for establishing social structures which ensured that labour was able to retain the full value of its input into the production process (Kitching 2012: 29). Ricardo's *Principles of Political Economy* was accordingly being read radically from the left by the 1830s and, in Ronald Meek's (1967: 71) words, the rise of a new economics in its wake "was determined by a belief that what was socially dangerous could not possibly be true". Yet it would be a mistake to think that the political project of the Ricardian Socialists died at Jevons's hands in the 1870s (De Vroey 1975: 431). By then, the cost-of-production theories against which Jevons oriented himself had successfully reincorporated the political interests of the privileged that the Ricardian Socialists had attempted to exorcise.

In response, Jevons certainly placed the individual at the forefront of economic theory, but this was not the conscious individualism stripped of social context that eventually became associated with the marginalist revolution (Milonakis & Fine 2009: 110). New negligibility, applicability and tractability assumptions (see Chapter 3) have been introduced into economists' market models since Jevons's day to wholly recast the idea of the individual to which individualistically oriented economic theory responds. The more precise the mathematical object that economists use to reason with, the less market models are constructed using techniques of Galilean idealization. As a consequence, the less real the individual inhabiting the models becomes. We are returned to the important difference between Marshall's and Robbins's competing conceptions of precision (see Chapter 1). Obviously, it is necessary to avoid ascribing to Jevons positions that were developed at a later date. Yet Marshall's definition, which linked precision with as close a match as possible between theoretical statements and supporting empirical evidence, was what Anglo-Irish economists generally understood by the word in Jevons's day. Despite his revolutionary intentions on other aspects of economic theory, here he had both feet planted firmly in the mainstream. His approach to mathematical market models existed entirely independently of Robbins's later recodification of the field around strictly logical precision.

Jevons's personal turn towards democracy was driven by a previous generation of economists' inability to recognize the public's increasing anxieties about ever more egregious forms of economic inequality (Sekerler Richiardi 2011: 94). Viewed from this perspective, the individual could not be reduced to logical propositions regarding the interplay of abstract demand and supply functions. It was necessary to treat them as a real person facing day-to-day dilemmas of how best to make ends meet if it were to be possible to break the spell of Millian individualism that served the interests of a deeply unequal social *status quo*. The underlying rationale of Jevons's economic theory was to squeeze the highest possible enjoyment out of available consumption possibilities (Groenewegen 2007: 249). Yet to understand this shift fully, the first thing to remember is that Jevons's consumer was not *merely* a consumer. Consequently, his hedonistic principle of pleasure-seeking economic agents extends beyond the enjoyment derived from the act of consumption to what has to be sacrificed in pursuit of that act (Black 1970: 30). The empirical features of the social context that economics imperialists typically treat as disturbing background noise is crucial to his market models. Jevons worked with a rounded coception of economic agency, an analytical ideal-type capable of being understood through careful statistical study of real people's lives, those who operated simultaneously on both sides of the market as consumer *and* labourer.

Jevons's (2013 [1871/1879]: 28–32) utilitarian commitments were evident in his use of the behavioural principle of a balance between pleasure and pain. He then translated the image of a mathematically tractable balance of forces (Mirowski 1991: 219) into an economically meaningful connection between the same individual's activities in the spheres of consumption and work (Maas 2005b: 638). Consumption might very well explain the purposive orientation of the labour process (Jevons 2013 [1871/1879]: 167), but that process imposes costs on the labourer in terms of physical depletion. The consumer's decision-making and the labourer's decision-making must therefore be treated holistically, for the very simple reason that they relate to the same person. To accept the consumer's experience as *the* quintessentially economic aspect – as economics imperialists do – is thus to rob the economic agent of an important part of their character.

Restoring both halves of the Jevonian agent is significant for the way in which the marginalist revolution, at least in its Anglo-Irish variant, might be read politically. In Joan Robinson's (1962: 66) terms, the utilitarian appeal to maximizing pleasure with respect to pain "seems to be heading straight for egalitarianism of the most uncompromising kind". What else could be read into Jevons's construction of the economic agent, she asks, than a warrant "to interfere with an economic system that allows so much of the good juice of utility to evaporate out of commodities by distributing them unequally" (Robinson 1962: 53)? Jevons was not arguing on the intellectual territory of the Ricardian Socialists, but nonetheless there were obvious egalitarian implications dotted around his work, even if they can be lost amidst the hectoring tone he often adopted in chastising members of the working class for their errant behaviour (Jevons 1883: 5–8). The crucial aspect of utility, he said, was that it was measurable (Jevons 2013 [1871/1879]: 49). The pleasure derived from consuming the commodity could therefore be quantified and compared directly to the pain involved in producing it. Critically, all distributional questions could be placed back in play by focusing on where within the social structure both the pleasure and the pain were concentrated. Who was in surplus, having exploited a position of social privilege to ensure that the utility they enjoyed in consumption was disproportionately large compared with the effort they had to put into the production of goods? And who was in deficit, receiving a measly utility return for the pain they had to put their bodies through while labouring? Statistical objects could be used to verify where the tipping point was between preponderant sensations of pleasure and preponderant sensations of pain. As the question driving mathematical market models subsequently changed from who did well out of the current social structure to whether the world within the model could be said to have reached equilibrium, these statistical objects were replaced by more abstract forms of mathematical notation (see Chapters 6 and 7).

Jevons undertook an experiment, which he published in *Nature* in 1870, designed to illustrate the marginal principle in relation to the body's aptitude for work. He showed what his intuition had already told him was true, which is that every additional increment of repetitive work is progressively more difficult for the body to withstand without replenishment. The pain experienced during the labour process increases in accelerated fashion the longer the working day (Jevons 1870: 159). However, the opposite effect was shown to operate for repetitive consumption. The pleasure experienced in each extra act of consumption fell off progressively compared to what had gone before. Taken together, these tests proved two things: first, that those who worked for a living failed to receive the compensation their efforts deserved; second, that the potential utility unleashed by modern mass production was largely squandered by placing life-affirming consumption possibilities in the hands of too few people.

This was a time, of course, of a factory system that required excessively long working hours for markedly poor rewards. Gratuitous consumption amongst the rich also took place in the context of mass deprivation elsewhere within society. Jevons's economics could speak directly to these issues to embrace the reformist zeal then in evidence in Britain, raising the possibility of addressing the distributional issues of the day. However, policy remained under the control of the purveyors of orthodoxy in Parliament, and they revealed no such appetite. It is perhaps unsurprising, then, that John Elliot Cairnes (2004 [1857/1875]: 25), the figurehead for the economics parliament still took to be authoritative, commented ruefully on "the repugnance, and even violent opposition" of the working classes to economic theory.

Jevons's attempt to make it more palatable to the masses invoked the combination of mathematics and number. His economic theory worked in its own terms through imagining a tractable mathematical structure propelled into being by hedonistic influences. Yet what brought this theory to life was the numerical demonstration that the pleasure which could be purchased in the act of consumption must always be mirrored by the pain that was experienced in the labour process. Pleasure counteracted by pain is something that people would have recognized from their own dual existence as consumers and workers. Jevons's market models require the presence of certain tractability assumptions, as is true of any mathematical model constructed with an explanatory purpose in mind (see Chapter 3). But in his case the mathematics is sufficiently simple that it does not overwhelm the potential realisticness of the model by enforcing accompanying negligibility and applicability assumptions that serve only to highlight how little the world within the model has in common with the world beyond. Even though Cauchy's metamathematical observations are not referenced anywhere in Jevons's published work, the emphasis on numerical demonstration that appears so vividly in his 1870 *Nature* article locates him squarely within such a tradition. Cauchy insisted on the priority of mathematical objects that had been found, rather than had just been proved in principle to be findable. Jevons also thought that mathematical reasoning tools must be plausible representations of actual economic experiences and not merely thinkable mental artefacts.

However, Jevons's place in the Anglo-Irish *Methodenstreit* is now typically seen only as a reflection of John Neville Keynes's eventual resolution (Lewin 1996: 1298), even though Keynes's position seems to be noticeably anti-Jevonian (Deane 2001: 138; Moore 2003: 5, 18). Despite setting out with the explicit goal "to avoid the tone of a partisan", Keynes managed to smuggle elements of the Austro-German *Methodenstreit* into his account (Keynes 1999 [1890/1917]: xviii, xiii). He acknowledged the significance of historical precepts to what he called the art of economics, but only having first changed their terms so that they could never be more than subsidiary to the real job of placing abstract economic theory on a secure logical footing (Colander 2001: 19–33). Jevons's search for a tractable mathematical structure for economic theory was very much in Keynes's category of the science of economics, whereas his numerical demonstrations help to prepare the scene for its art. Keynes's subordination of the art to the science undermines the holistic nature of Jevons's approach by significantly downplaying the Cauchian metamathematics of numerical demonstration. Jevons (1876: 627, 624) would countenance no such restriction in his vision for economics as an "aggregate of sciences ... The fact is it will be no longer be possible to treat political economy as if it were a single undivided and indivisible science". Keynes's *Scope and*

Method thus casts Jevons adrift in a metamathematical no-man's land not of his own making (see Chapter 2).

We are thereby returned to Jevons's disagreement with Leslie. This was over whether economic knowledge historically derived should supplement or supplant economic knowledge logically derived. Leslie was a supplanter, reacting to the lack of social awareness in the dominant abstractions of contemporary theory to mobilize for the wholesale deletion of deductivist methods from economics (Tribe 2007: 227). Jevons, though, was a supplementer, aiming to reinvigorate deductivism by using historicist techniques to ensure that theoretical abstractions captured the essence of actual lives (Henderson 1994: 506). Jevons embraced logical argumentation, but not to the extent that logical precision could replace empirical precision. Leslie's approach, said Jevons (1876: 623), was "absolutely essential" to the operation of a fully rounded economics: "I cannot easily conceive any more interesting or useful subject of study than that which Professor Leslie advocates". Yet even though they appear alongside one another as equals in Jevons's vision of economics, still the only way in which he could announce their equality was through arguing that the abstract and the concrete were to "naturally be divided" (Jevons 1876: 624). They related to equally important intellectual endeavours, but nonetheless were separate. As this is so significant for what comes next in the prehistory of economics imperialism, the roots of Jevons's case for subdivision *within* economics must be further explored. Attention has to turn to his embrace of the nineteenth-century Anglophone psychophysiological tradition. This is the origin of the filling instructions for his general argument pattern, but he clearly struggled to render them anywhere near fully concrete.

JEVONS AND THE PSYCHOPHYSIOLOGICAL TRADITION

It is impossible to get an all-round appreciation of what Jevons's *Theory of Political Economy* meant for economics without situating its contents within Victorian Britain's discussion of the science of mind. Economics imperialism is usually understood against the backdrop of a largely teleological process that created a separate and distinct science of the economy (Mäki 2002: 238; Fine & Milonakis 2009: 40). Yet any attempt to anoint Jevons as the founder of an autonomous economics appears to be misplaced. His efforts to build a general economic theory proceeded on the basis of a motive-laden economic agent, but the motives themselves were imperceptible using economic theory alone (Jevons 2013 [1871/1879]: 33). Here we see, right at the very outset of his theoretical framework, Jevons (1876: 623) attempting to distance himself ontologically from what he called "the fallacy of exclusiveness": the idea that economic models could stand on their own, even as purely mental artefacts. The source to which economists had to look, he suggested, was the evolving nineteenth-century debate about free will and the distinction between mind and matter. Not without good reason, then, Margaret Schabas (1990: 125) has argued that, for Jevons, the destiny for economics was for it to become a subset of psychology.

Some important distinctions must first be discussed. In the 1930s, Lionel Robbins (1984 [1935]: 22, 1936: 171) emphasized the logic of choice under conditions of scarcity as his approach to a subjective theory of value, citing Jevons as his inspiration (see Chapter 5). However, Robbins saw no need to comment on the historical patterns of socialization through which the scope of feasible choices was set, but Jevons did. Or perhaps more accurately, Robbins (1984 [1935]: xxxvi) contracted out all such need to an altogether different mode of study he called political economy, whereas Jevons (1876: 624) maintained it within the different internal categories of economic theory. Jevons (1876: 619) wanted to create an economics fit for an age of middle-class sensibilities about the structurally inscribed limits of the choices available to the less fortunate, and to this end he treated as inseparable choice and the social factors underpinning choice. His analysis of motives is complex, and at times perhaps more than a little confused, as befits its place within nineteenth-century studies of the mind that oscillated between emphasizing the generically psychological and the more specifically psychophysiological determinants of action (Kreisel 2012: 27). But nowhere in Jevons's work can one detect what subsequently became the feature of mathematical market models that helped to propel economics imperialism: motiveless agency enacted simply to achieve the technical objective of maximization and thus to render the model determinate.

For Jevons (2013 [1871/1879]: 10), economics had to be based on behavioural psychology if its discussion of motives was to succeed in the overall objective of "comparing quantities of advantage or disadvantage". It should be immediately obvious that in talking about "comparing quantities" Jevons was emphasizing the need to make economic arguments through the use of number. It was within such comparisons that the balance could be determined between the enjoyment an individual could extract from the commodities produced through the world of work (Jevons's "advantage") and the labour that same individual had to contribute to the production process (Jevons's "disadvantage"). Yet these comparisons were acts of the human mind, pure and simple. If it was impossible to derive plausible explanations for how the mind prompted the body into action, then the very goal of genuine economic understanding would forever remain off limits.

It is here that Jevons displayed his foremost intellectual debt to Richard Jennings's little-known 1855 book, *Natural Elements of Political Economy*. Indeed, "little-known" is almost certainly an understatement, because pre-Jevons it was pretty much entirely ignored and post-Jevons it has long since slipped back into obscurity. Jennings's economics looks very unusual today, with its references to the sensorium (the part of the brain responsible for receiving and interpreting sensory stimuli) and to the afferent and efferent trunks (the carrying capacity of the nerve fibres of the central nervous system). Beneath this obscure Victorian language, though, there lay a number of general assertions that heavily influenced Jevons. Jennings (1969 [1855]: 21, 16) argued that "elementary principles of human susceptibility and of human action" can be unlocked only through engaging the science of mind. These are the origins of the actions that economists should wish to study, and they "cannot be thoroughly understood until the subordinate principles of psychology have been adequately investigated". If ever somebody else had written a manifesto for Jevons's economics, then this was it.

In this reading, mathematical market models contain no explanatory content of their own; this arises only when statistical objects corroborate the underlying psychological theory. Jennings (1969 [1855]: 99) believed that all laws of human behaviour reflected functional "increments of sensation", which enabled them to be described using a combination of mechanical metaphors and enumerated evidence. Jevons was familiar with such a way of thinking from his prior readings in the natural sciences, and his diary entries from 1860 record the almost instantaneous shift in his thinking on value coinciding with his first exposure to Jennings's work (White 1994: 205). Jennings (1969 [1855]: 24, 18) insisted that "the conception of value mentally entertained" could proceed only by means of "an inquiry into that universal relation of mind to matter in which originate a large proportion at least of the sensations, the conceptions, and the actions of mankind". The only point on which Jevons baulked was at the thought that physiological activity picked up by the sensorium was enough on its own to explain why the conduct of economic agents could be captured by the underlying principles of marginalism (Gallagher 2006: 124).

For Jennings (1969 [1855]: 142), "the mystery of action is solved" only when combining the behavioural laws that apply respectively in the fields of consumption and production. Consistent with everything else in the *Natural Elements*, these laws appear to be purely physiologically based; his psychophysiological theory had much less psychological content than Jevons's. Consumption is to be understood, from Jennings's (1969 [1855]: 73) perspective, as "[t]hat action and reaction of matter and man, by which matter supplies the means of gratification to man, while man diminishes or annihilates the valuable properties of matter ... [D]uring its continuance, the operation of the *afferent* trunks of nerve-fibre prevails". The afferent trunks conduct impulses to the central nervous system. Production, meanwhile, is to be understood as "[t]hat action and reaction of man and matter, by which valuable properties are imparted to matter, whilst reflex impressions of resistence are felt and sustained by man ... [D]uring its continuance, the operation of the *efferent* trunks of nerve-fibre prevails" (Jennings 1969 [1855]: 73). The efferent trunks conduct impulses from the central nervous system.

Jennings's appropriation of physiological explanation produced something that had really not been seen before in economics (White 2002: 13). However, Jevons was able to show that the structure of Jennings's thought was logically separable from its contents. The structure enabled the mathematization he sought for his account of market models, whereas the content was incompatible with what would set in motion the internal dynamics of the mathematical objects he would use to reason with. Jennings reduced all economic behaviour to the operation of conducting impulses from the central nervous system in his bid to depict a functioning economy as the balance between effort expended in production and enjoyment obtained from consumption. Jevons maintained a focus on that balance but dispensed with all of the surrounding physiological language, basing his general subjectivist theory of the economy instead on a hedonistic psychology of utility and disutility. Here, at last, the mathematical shell from the 1862 British Association for the Advancement of Science paper could be filled in.

There was little room in Jennings's *Natural Elements* for behaviour to be other than the routine operation of something akin to a gravitational force that required an automatic response from the body (White 1994: 198). In some instances he got close to admitting to the presence of volition as a background event (Jennings 1969 [1855]: 10, 22, 125, 135), but it is clear that these are offered very much as exceptions rather than the rule. Indeed, he argued that such instances of will-governed behaviour take place "much less frequently than is commonly supposed" (Jennings 1969 [1855]: 132). We might have good reason for wanting to look upon our own actions as if we were taking conscious decisions at every step along the way, but for Jennings this is to overstate the extent of agential autonomy. The sensations upon which the economist might be able to build a general theory are "easily recognised as … being distinguished from other mental states by the characteristic that [they are] *immediately* caused by the actions of the bodily organs, without the occurrence of any intervening mental state" (Jennings 1969 [1855]: 83, emphases in original). In other words, whenever one observes economic action it is much more likely than not to be purely reflex behaviour (Zouboulakis 2014: 32).

Jennings was by no means alone in attributing behaviour to a passive sensationalism, as this tapped into a rich vein of thought in Victorian Britain. As Harro Maas (2005b: 621) notes, "general man-machine comparisons … were in vogue". Nineteenth-century British advances in the science of mind tended to conclude – somewhat paradoxically – that the mind was merely epiphenomenal of the body (Smith 1981: 164). An increasingly sophisticated understanding of the reflex system resulted in ever more autonomy being attributed to centres of innervation, through which the excitation of nerves takes place (Jacyna 1981: 112). This argument was then put to work in various ways, including to explain the persistence of the social institutions that the economics profession had recently come to criticize. Henry Maudsley's *Physiology of Mind*, for instance, published in the same year as Jevons's attempts to reconcile theoretical and historical economics, was positioned against the existence of an immaterial ego that could weigh competing visions of social organization (Maudsley 1876: 63). A prior chain of psychophysiological cause and effect had to underpin all behaviour, according to Maudsley (1876: 412), otherwise human conduct would dissolve into "an unmeaning contradiction in terms and an inconceivability in fact". Viewed from this perspective, the social institutions which project throughout the population the conventions of everyday morality themselves depend upon a predictability of conduct that is at odds with free will.

These developments in nineteenth-century British physiology began with the pioneering work of Marshall Hall in the 1830s. Hall (1837: 50) questioned the role of consciousness in human decision-making, situating behaviour at the level of "excito-motory" acts that arose beyond the need for consciousness. Thomas Laycock (1845: 302) went further, describing situations not merely in which there was no need for consciousness but in which there was reason to question its very existence. He believed it to be a coincident phenomenon of the impact of neural organization, whereupon the automaticity of even the brain's higher functions casts doubt upon the whole sphere of volition. But nobody went as far as Thomas Huxley (1874: 200) in openly declaring the illusionary nature of feeling and willing. The human agent disappeared entirely in this view, replaced by the image of "man as automaton" (Maas 2005a: 155). The title of Huxley's 1874 essay, "On the Hypothesis that Animals are Automata", perhaps says it all. Man-machine metaphors had given way in many quarters to the simple assertion that the human being could be thought of *as* a machine, whereby consciousness was manufactured by the workings of the body (Huxley 1874: 212).

Jevons's economics should be considered in relation to these debates, because it was conceived in their midst, even as it stood apart from their conclusions. This was much further apart, it should be noted, than his successors in phases 2–4 of the prehistory of economics imperialism. They showed no interest in the psychophysiological debates that interested Jevons, but nonetheless got close to endorsing man-as-machine metaphors through their use of mathematical logic and mathematical expression. For Jevons, though, the emphasis on reflex action quite clearly posed difficulties for any discussion of the propriety of conduct. The more that volition was written out of the explanation, the less possible it was for economic theory to retain its established position within the moral sciences. Jevons (1876: 624) was evidently attracted to the idea that emulation of the natural science methodology of experimentation could lead economists towards universal laws of behaviour in which volition might begin to look a lot like the choice to follow rules. Yet at the same time his scientific writings display an epistemological commitment to fallibilism (Hutchison 1982: 375). Economic laws were only ever likely to be tendentially true, he wrote (Jevons 2014 [1874]: 792), in contrast to the literal truths that continued to be claimed by the old guard (see, e.g., Cairnes 2004 [1857/1875]: 56). For Jevons, economic models could never reproduce the complexity of actual economic systems, because it was impossible to state behavioural laws that would be determinate for all people at all times: "laws and explanations", he wrote, "are in a certain sense hypothetical, and apply exactly to nothing which we can know to exist" (Jevons 2014 [1874]: 458).

However, Jevons's "applying exactly to nothing" is rather different to how the same condition is typically discussed today regarding Sugden's credible-worlds methodology for economists' hypothetical mathematical models (see Chapter 3). There, a complete disconnect with the world beyond the model can be tolerated as long as the world within the model is thinkable (Sugden 2000: 24, 2009a: 17). For Jevons, though, there were limits to this distance based on the need for the two worlds to bear close resemblance to one another. We are thus returned to the significance of the division of labour that Jevons posited *within* economics and one that seriously undermines the notion of a linear prehistory of economics imperialism. Economic theorists could only present their abstract laws as being of the most general kind, while exact forms of economic relationships were to be constructed by historical economists using empirical analysis sensitive to the specificity of the surrounding institutional environment (Flatau 2004: 79). The existence of free will always raised the possibility of important differences between the general and exact forms of economic relationships (Porter 1986: 177).

JEVONS AND THE NINETEENTH-CENTURY FREE WILL DEBATE

The blurring of the mind/matter distinction in nineteenth-century British psychophysiology presented Jevons with a dilemma. It certainly helped to render economic theory susceptible to styles of thought that had long since dominated the natural sciences, most obviously the mathematical approach to mechanics that had facilitated James Clerk Maxwell's successful scientific unifications in the 1860s. But at what cost (see Chapter 1)? Maxwell's accomplishments were hot news throughout the period when Jevons was initially refining his theoretical system, raising his faith that the development of mathematical market models along Maxwellian lines could help circumvent the currently "chaotic" state of value theory (Jevons 1876: 620). Yet how might he include a passably realistic account of economic agency in the relevant system of equations? At that time, John Stuart Mill was working with an introspective method he said could reveal first principles of economic behaviour by search-ing deep inside the self (De Marchi 2002: 311). An inner eye could penetrate one's own mind, distinct from the outer eye helping to make sense of other people's interaction with the matter of the external world (Maas 2005b: 626). Jevons (1877: 178) dismissed Mill's appeal to "the private laboratory of the philosopher's mind" – "What a convenient science!" – and hoped that the experimental methodology of psychophysiology would offer a more convinc-ing alternative (Daston 1978: 196).

Jevons's correspondence from the early 1860s shows that he was clearly aware of the ethical difficulties involved when the science of mind embraced strict physiological accounts of human behaviour (White 1994: 221). However, the only other aspect of his studies that was keeping pace with the develop-ment of his economic theory at this time was in the field of logic. While retain-ing much of the basic structure of Jennings's economic arguments, Jevons struck out on his own by replacing the physiological content with elementary mathematical reasoning tools that reflected more what he had learnt in logic classes than in mathematics classes. It is one of the most frequently quoted lines from Jevons, taken from the preface to the second edition of his *Theory of Political Economy*, that "the mathematical treatment of Economics is coeval with the science itself" (Jevons 2013 [1871/1879]: lxiii). Yet the significance of Jevons's adoption of mathematics lies less in what it enabled him to say in its own terms than in what it allowed him to avoid importing from more phys-iologically oriented authors in the psychophysiology tradition. After all, the mathematics being invoked was little more than straightforward arithmetic placed within a primitive algebraic outer layer. It was neither necessary for fleshing out his hedonistic calculus of the balance between pleasure and pain, nor a means of establishing the existence of free will.

The text of Jevons's *Theory of Political Economy* reveals the influence of William Carpenter, a mid-nineteenth-century physiologist. Carpenter was one of the first to detect that the trend in psychophysiology was to depict in puppet-like form what should really be "the Thinking Man" (Carpenter 1852: 510). He concerned himself instead with retaining the idea of self-determining human agency, but even so felt it necessary to concede to his more physiologically oriented colleagues something spontaneous in the character of individual volition. For Carpenter, the task of the will became to select amongst available options that reflected the broad range of neural capacities, rather than to follow any given option without due reflection (Daston 1978: 203). Carpenter borrowed the concept of physical forces to posit analogous mental forces: just as physical forces propel matter to behave in particular ways that eventually become recognisable as patterned action to the observer, so too mental forces propel the human body to behave in similarly patterned ways (Carpenter 2013 [1855]: 799). The mental forces were themselves unobservable, but the effects they created via the action of the body lent themselves to rather straightforward forms of empirical study (Smith 1977: 220).

Significantly, they were also consistent with a type of mathematical expression that made it relatively straightforward to reduce human action to a series of equations. The presumed likeness between mental and bodily states presented Jevons with the loophole in prevailing psychophysiological arguments he was looking for. It enabled him to assert the presence of free will without ever needing to demonstrate the basis for that assertion, and it also opened economics to the type of experimentation that would take it forever past Mill's introspectionism. Numbers were thus to rule the day, in a seeming invocation of what Hilbert was later to call concrete axiomatization (see Chapter 2). The numbers through which Jevons sought to reconstruct economic theory would only be genuinely meaningful in the presence of corroborating evidence drawn from real-life situations. Economics in its Jevonian mode was thus still significantly shy of later metamathematical claims that mathematical theories – including mathematical theories of basic market models – could be rendered true in their own terms.

Jevons was left merely to look for suitable mathematical techniques to enable him to build the resulting economic theory on strictly logical foundations grounded in number. He found them in the work of the Belgian statistician, Adolphe Quetelet (Inoue 2012: 49). Quetelet's particular brand of positivist empiricism – the search for law-like properties within observable human behaviour – became extremely popular in Britain in the 1860s as a new front in the mind/matter debate (Szreter 1996: 170). Jevons was evidently an early convert, purchasing in 1857 Quetelet's *Sur l'Homme et le Développement de ses Facultés, ou Essai de Physique Sociale* (albeit in the English version translated as *A Treatise on Man and the Development of his Faculties*). He was feeling sufficiently confident in his use of the text to tell his sister the following year that he was engaged in a project that echoed Quetelet's "science of man" (Mosselmans 2007: 34). This was four years prior to presenting his formative mathematical study of the essence of economic theory to the British Association for the Advancement of Science. In the intervening period, Quetelet's study of the logic of probability provided a constant lens through which Jevons familiarized himself with the economic theory of the day (Groenewegen 2003: 159). The free will problem disappeared within Quetelet's (1842: 96–106) averaging techniques. Whether one sees Jevons's appropriation of probability theory as a solution to that problem or as a convenient dodge, it nonetheless gave him more stable foundations for developing his market models as algebraic shells encasing predominantly statistical objects.

Quetelet had pioneered the method of selecting a single representative numerical value that could be considered sufficiently of the type to stand in for a whole host of disparate social observations (Duncan 1984: 108). Jevons's *Principles of Science* acknowledged the influence on his economics of Quetelet's study of the "comparative frequency of divergence from an average" and, in particular, the symmetry that can be assumed to exist within the frequency distribution (Jevons 2014 [1874]: 188). The representative values on which Quetelet focused could therefore be mapped onto something like the modern-day bell curve, as formalized by Carl Friedrich Gauss in the 1820s. Jevons appears to have been transfixed by the idea that the imprints of natural order could be detected in the pattern of deviation from behavioural norms, deviations which in statistical terms could be explained away as errors (Porter 1986: 176). Quetelet might well today be thought to have exaggerated the regularity of human behaviour so that it would fit clear statistical functions (Ambirajan 1995: 206). But his argument that a sufficiently large number of observations would show that human behaviour is normally distributed allowed Jevons to disarm philosophical objections to contemporary developments in psychophysiology. A statistical trick could be used to answer an ethical question. Economic agents could be allowed as much free will as they pleased if it could be safely assumed that will-governed deviations from the norm in any case cancel one another out.

Quetelet followed Pierre-Simon Laplace's assertion that, in the natural sciences, apparently arbitrary events are not down to chance *per se*, so much as to the limits of human knowledge (Lee 2012: 22). For Jevons, it was a basic ontological commitment that the actions of lone individuals did not permit the observational basis for the derivation of general laws, because any single individual's actions had an unknown propensity for appearing to have been affected by chance (Peart 1996: 236). Whereas Mill had argued that access to *each* individual's mind was the source of understanding law-like economic behaviour, Jevons was concerned that the representativeness of *any* individual's mind could not be taken for granted (Udehn 2001: 51). Quetelet's normal distribution provided a route out of these concerns, because it rendered redundant the individual as a source of empirical data, replacing the promptings of any single mind with that of a typical mind. This enabled Jevons to defer to Leslie's historicism when attempting to explain the manifestation of institutionally embedded forms of economic *life*, but at the same time to claim that, as a matter of economic *theory*, individual behaviour tends to reflect "accidents" that neutralize each other on average. Treating the population as being large enough to describe the properties of a normal distribution would be sufficient on its own to ensure that the representative criterion would hold for the mean of the distribution (Mosselmans 2007: 36). Jevons's commitment to ensuring the presence of the real world within his market models thus seems less secure, unless it can be known for sure that will-governed choices of actual economic agents are normally distributed in practice.

Quetelet's *l'homme moyen* – the "average man" – reappeared consistently throughout Jevons's theory of exchange (Jevons 2013 [1871/1879]: 75–166), providing the agential mechanism through which market outcomes are activated (Zouboulakis 2014: 29). The law of large numbers is used as an offset for disturbing causes rooted in unpredictable will-governed behaviour, whereby "the market" can be understood as an abstract phenomenon distinct from actual markets (Watson 2018: 6–9). However, the difference between the two is accentuated by the fact that Quetelet's average man construction appeared in many different guises in his work as a whole. It played to Jevons's early interest in mathematical logic that it originally expressed only numerical relations between statistical data, but it also played to his later interest in the historical underpinnings of lived market relationships that Quetelet paid increasing attention to the possibility of competing average *men*, each one specific to a particular time and place (McCann 1994: 152). Perhaps unsurprisingly, then, *l'homme moyen* appeared to oscillate in Quetelet's subsequent work between statistical artefact and something that should be treated as if it was real (Hacking 1990: 108). Hilbert's later distinction between concrete and formal axiomatization (see Chapter 2) is visible here in pre-emptive form, but only when trying to separate two entities that Quetelet often treated interchangeably.

It is generally agreed that Jevons failed to solve the difficulties associated with the application of probability in empirical work (Ambirajan 1995: 202). Phenomena of the mind were simply too complicated for probability to capture in exact form (Jevons 2014 [1874]: 177). He never attempted to quantify the uncertainty of his estimates, relying instead on "a quick ad hoc fit" that retained the shape of the probability distributions (Stigler 1999: 86). This is almost certainly all Jevons needed to understand his mathematical market models in their own terms. But it also invokes a slippage in the concept of number to which he referred. His metamathematical commitment to using numbers in the concrete axiomatic sense of being supported by corroborating evidence appears to have strengthened over time. Yet in his mathematical applications he seems to have increasingly used numbers derived from statistical theory as if they had actually been derived from observations of society. As will become evident in later chapters, such a shift was crucial in the developing story of economics imperialism, as mathematical market models increasingly reflected economists' conceptions of credible worlds rather than the real world.

It is important in this regard that Jevons tried to enforce a new standard in probability theory by which the word "mean" would be reserved for empirical observations of actual people going about their daily business, leaving the word "average" to be used solely in connection with statistics (Duncan 1984: 109). The average man construction he took from Quetelet's *l'homme moyen* was not, and was never intended to be, a "mean man". In the final reckoning, Jevons's market agents were anthropomorphized elements of a probability distribution and therefore not anything that should be mistaken for real people. This was despite his evident concern for propelling public policy into a much deeper understanding of how real people lived their lives. With his market agents reduced to statistical artefacts and given a dynamic persona through his attachment to hedonistic psychology, all that remained for him to do was to describe the essence of their economic character. It was entirely in keeping with his approach to abstract questions in general that he also found this answer in his ongoing study of logic.

The primitive that runs throughout Jevons's theory of logic is the fundamental notion of equality (Schmidt 2011: 545). His 1869 treatise, *The Substitution of Similars*, argued that the basic mode of mathematical reasoning follows closely the everyday practice of rendering one thing knowable through its likeness to something else that is definitely known. The image of explanatory unification is not far from the surface here, but Jevons did not head in that direction. Instead, he was solely concerned with what the imaginative act of swapping one thing for another meant for his study of logic. "[W]e continually argue by analogy from *like to like*", he suggested (Jevons 1869: v, emphases in original), where "the instrument of substitution is always an equation" (Jevons 1869: 19). The mind, for Jevons, is culturally conditioned to turn inequalities into equalities as a means of facilitating understanding of that which is not yet fully known. As such, all economic activities in the abstract market realm inhabited by Quetelet's average man must be directed towards the production of an equality, the stationary point in the hedonistic calculus that respects both the pleasure derived from consumption and the pain caused by labouring. Jevons thus took an extremely circuitous route from psychophysiology to logic in his search for the fundamental underpinnings of his economic theory. At some stage in that journey, in order to arrive at his destination where the notion of equality could provide the intellectual scaffolding for the rest of his economic theory, he appears to have wavered when it came to his own ontological commitments. Was it the task of economists' mathematical market models to capture the lives of real people or of a hypothetical economic agent? This question re-emerges in each of the subsequent phases of the prehistory of economics imperialism, where the answer seems to lean ever more towards the latter.

CONCLUSION

When viewed at a distance, Jevons can appear to be an all-important trailblazer in the prehistory of economics imperialism. He was certainly a forerunner of attempts to turn economics into a modelling science, and he was eager to mathematize his market models. However, he flatly denied the related claims that economic theory could stand alone from other forms of social enquiry and that such separation bequeathed a superiority that justified the colonization of neighbouring fields. Jevons ultimately seems to have put his faith in the existence of a stochastic world in which the outcomes of will-governed behaviour describe a well-ordered probability function where individual points are nonetheless randomly assigned. His latent metamathematical preferences suggest that this faith had more to do with modelling expedience than ontological commitment, but still his statistical objects required the embrace of a particular behavioural psychology if they were to be activated in an economically interpretable manner. The narrative content accompanying Jevons's rudimentary mathematical market models is suffused with a hedonistic worldview, reducing the task of all abstract economic theory to "tracing out the exact nature and conditions of utility" (Jevons 2013 [1871/1879]: 43). Explanatory unification therefore took place in his case on psychological rather than economic terms. As a result, Jevons looks to be more an antecedent of Ralph Souter (the originator of the term "economics imperialism" but an advocate of friendly cohabitation amongst the social sciences) than of Anthony Downs, Gary Becker, Duncan Black, James Buchanan, Gordon Tullock, Mancur Olson and Richard Posner (the innovators in early proto-economics imperialist endeavours) (see Chapter 1).

A straightforward corrective thereby arises whenever it is suggested that Jevons had been pointing towards the future that eventually transpired. Today's mathematical market models have very different content to his, because they are founded upon very different traditions of thought. No modern economics textbook would devote time to Jevons's struggle to find a plausible position within the free will debate. His engagement with nineteenth-century psychophysiologists has been lost from mainstream disciplinary history. Neither should we expect students to witness their lecturers reflecting on whether Jevons's approach to number survived his own forays into the logic of statistical averaging. These were obviously very important questions for the success of his theory, but they rapidly fade from view if Jevons is repackaged – as he typically has been since Keynes's 1890 *Scope and Method* – to promote the idea of smooth linear progress from past to present. My hope is to restore these missing historiographical links as a means of getting to the heart of the prehistory of economics imperialism. This is crucial to any subsequent adjudication on the soundness of its foundations as an explanatory activity.

The prehistory of economics imperialism is always as much about what was taken out of economic theory as the content that was added to it. As is made clear by Keynes's role in mediating how Jevons's legacy has been understood, this is often what was removed when others tried to establish after the fact a canonical interpretation of important analytical changes to the market model. Jevons stands out in my historical chapters in this regard, due to the frequency with which he asked himself the question of what counted as an excision too far. Yet instead of using this as a reason to row back from his theoretical developments, it tended to lead only to sources of equivocation in his writing. It is difficult to resist the feeling that Jevons never quite made up his mind about where he stood on many of the biggest analytical issues. This allowed others to subsequently speak in his name, to cure him of his hesitance by imposing the meaning on his work that they most desired it to have.

Jevons's economic theory revolves around a clearly abstract economic agent, and his use of the differential calculus to explain the balance between pleasure and pain seems to locate its most obvious hypothetical characteristics in a strict instrumental rationality that makes possible the calculation of how each prospective activity will add to the sum of pleasure with respect to pain. But the related use of statistical averages when populating his models bypasses the need to say anything about the rationality on display. Each person can be allowed to conduct themselves in whatever way they choose as long as deviations from the norm cancel one another out to leave *l'homme moyen* focused only on net utility considerations. Jevons emphasized the need to talk about real lives being led by real people if economic theory was to recover from the very low regard in which it was held by the public, but real people are the one thing that his approach to mathematical market models leaves no room for. Despite the complimentary tones with which he described Leslie's historicist economic methodology, it is impossible to start with Leslie's attempt to account for time-place specific modes of rationality but end up with Jevons's theory. There is maybe little wonder that other people have instead interjected their own views on "what Jevons really meant" in the presence of such ambiguity.

Nonetheless, Jevons remains central to my story because of his insistence that it was necessary to work outwards from the thoughts of the psychophysiologists to show how economic theory might be reset on abstract marginalist principles. However, a clear tension runs through his attempt to say, as per the ethicists of his day, that economic agents must be granted free will but also, as per the logicians of his day, that this free will anyway led to predictably distributed outcomes. It required him to depart from his own commitment that the world that had produced unconscionable extremes of inequality should be adequately represented in his mathematical market models. He thus flirted with the unstable interstitial space between empirical precision and logical precision for the purpose of constructing his market models as statistical objects. Jevons's successors in the prehistory of economics imperialism allowed for causal content to be further bypassed in an obviously empirical sense, in favour of a purely logical treatment of decisions that were taken at the margin. Over time, the mathematization of market models changed register under the influence of competing metamathematical positions.

Jevons's economic theory is nothing if not mathematical, but the form of mathematical expression he favoured focused heavily on number. These were numbers that ultimately relied on new statistical techniques of averaging rather than the experimentation he said he preferred methodologically. But in principle at least they were fully susceptible to corroboration through observation of real-life situations. This is what Cauchy argued was necessary for a mathematical expression to simultaneously encompass a truth claim (see Chapter 2). Yet marginalist techniques also enabled economists' underlying market models to be mathematical in altogether different ways, including at the hands of those who were striking out into a post-Cauchy world that remained a metamathematical unknown at the time. I now move on to assess the contribution of Lionel Robbins to the prehistory of economics imperialism. Nothing becomes any less complicated in the following chapter, because Robbins appears to have had one foot in each of the Cauchian and post-Cauchian metamathematical camps, while explicitly positioning himself against the whole trend of mathematization in economics.

CHAPTER 5

Economization as proto-imperialist move 2: Lionel Robbins and the search for an analytical definition of economics

INTRODUCTION

For someone always seeming to doubt his own originality, Lionel Robbins had a considerable effect on economists' professional self-image (Backhouse & Medema 2009b: 810). The very first paragraph of his autobiography asks his reader to think of him merely as someone who repeatedly found himself in the right place at the right time to comment on what others were doing to reshape economic thought (Robbins 1971: 11). The Preface to the first edition of his most famous work, *An Essay on the Nature and Significance of Economic Science*, strikes the same tone. It positions him as simply rendering increasingly legible analytical themes that were the creation of other, more visionary economists (Robbins 1932: xlii). This chapter focuses primarily on Robbins's definition of economics as the study of choice under conditions of scarcity, because it provides another important turning point in the prehistory of economics imperialism. Even in relation to his most celebrated achievement, though, Robbins said that he was merely doing other people's intellectual bidding for them, in particular that of the English marginalist pathfinders, Stanley Jevons and Philip Wicksteed (Robbins 1984 [1935]: 22). At most, he allowed himself credit for seeing more clearly than anyone else what united the otherwise disparate advances of those he considered to have reworked Jevonian and Wicksteedian insights most effectively. He always held the promise of scientific unification in exceptionally high regard.

Robbins has been taken at his word by other economists, as he is today best remembered as the foremost theoretical synthesizer of his day. More attention was being paid in the 1930s than previously to trying to bridge the gap in Anglophone economics between the English marginalists and Carl Menger's Austrian followers (Vaughn 1994: 14). Robbins was well placed to succeed in this venture, being fluent in German and therefore able to read for himself countless works that were yet to be translated into English (O'Brien 1990: 162). He also used his position as Head of the Department of Economics to bring to the London School of Economics Austrian economists he had previously befriended during his visits to Vienna (Robbins 1971: 91). Robbins revelled in the eclecticism afforded by such disparate influences, allowing him to be, in the words of his one-time LSE colleague, Richard Lipsey (2009: 846), "such a superb popularizer".

The equilibrium concept was one aspect of economists' theoretical instruments that was subjected to the Robbins treatment (Endres & Donoghue 2010: 558). No single national tradition of economic theory could be said to own that concept, and by his day it had been generalized far beyond its use by the early marginalists (Martin 2009: 520). Accepting Robbins's famous scarcity definition ensured that all subsequent scholarly activity had to coalesce around the equilibrium concept. When each new cohort of students has taken its first steps within the subject field for the last 70 years, it has been told that economics reduces to the study of choice situations under conditions of scarcity. The Robbins definition, in other words, has become *the* definition of economics, deepening professional socialization to the equilibrium concept along the way (Davis 2005: 191).

Robbins's definition thus worked as intended to sideline the Cambridge tradition that developed around Alfred Marshall's *Principles of Economics*. Marshall's use of marginalist techniques spoke directly to the classical economists' concerns for harnessing production to solve distributional questions (Reisman 1990: 84). Robbins did not believe that this was the job of the economist. Instead, he provided the first tentative steps towards the eventual formal axiomatization of the subject field along Hilbertian lines, as his scarcity definition increasingly emptied the equilibrium concept of its empirical content (Backhouse & Medema 2009a: 468). Greater distance was henceforth placed between the world within the model and the world beyond. In narrowing what counted as economics to an abstract feature of a hypothetical model world,

Robbins also expanded the range of social experiences that were susceptible to economists' techniques. His core contention was that economics could *only* be about the efficiency of action in situations where one choice necessarily precludes another. Wherever it is possible to identify a constant event conjunction in which choice-oriented behaviour is conditioned by the contextual factor of scarcity, Robbins's technique seems to apply. Such situations are by no means restricted to those conventionally associated with the substantive realm of the economy. As Márcia Balisciano and Steven Medema (1999: 282) note, his scarcity definition is therefore an important part of the explanation for why "[t]he economist in the modern era is often pictured as the bull in the china shop, taking his cold, unfeeling benefit-cost analysis here, there, and everywhere".

It thus seems quite fitting that his friend, William Baumol (1984: vii), has argued that Robbins's *Essay* "can only be regarded as a dangerous revolutionary document". It might be difficult today to find any economist who expressly identifies themselves as a card-carrying Robbinsian, but his influence has been at once both more subtle and more profound than producing conscious devotees. It is not necessary to explicitly acknowledge a lineage to Robbins to be following in his footsteps. Adherence to orthodox economics methodology revolving around choice-theoretic maximization problems is almost always enough on its own. As such, Robbins certainly deserves his place within the first rank of economics methodologists. But this is for both good and ill: for the additional safeguards he brought against the imprecise specification of economic problems, but every bit as much for what he tried to remove from economics in the search for a value-free subject field that could parade its ostensible neutrality between ends (see Robbins 1971: 147, 1979: xviii, 1984 [1935]: 24). The constant refrain from economics imperialists that their maximization models allow them to concentrate on purely objective research has its origins in the *Essay*'s second chapter, "Ends and Means". However, Harro Maas (2009: 515) has argued that, "[a]fter Robbins, most economists fell into a kind of Kantian slumber", "mere engineer[s] of maximisation procedures", according to Fabio Masini (2009: 421). Nobel laureate Amartya Sen (1980: 363) has complained that the success of the *Essay* "kept prescriptive studies somewhat immersed in a pool of apologies".

The chapter now proceeds in four stages in an attempt to show that Robbins should be seen as the standard bearer for the second phase in the prehistory of economics imperialism. This is the proto-imperialist move I call "econ-omization". In section one, I outline the delimitation strategy embraced by Robbins in signalling which activities should capture economists' attention. It is as important for what follows to concentrate on what his scarcity defi-nition placed outside economics as on the territory it claimed as its own. The reduction of economic theory to studying the abstract act of econo-mizing eventually made economic analysis ripe for taking over other social sciences. But from where did Robbins get that idea? In section two, I show that potentially promising leads ultimately come to nought when attempting to locate Robbins's formative influences in either the Austrian School (via Menger, Böhm-Bawerk and Wieser) or the Lausanne School (via Walras and Pareto). In section three, I show that the same is true of the Anglo-Irish mar-ginalism of Jevons and Wicksteed, whatever Robbins might have said about his "especial indebtedness" to Wicksteed in the Preface to the *Essay*'s first edition (Robbins 1932: xlii). The prehistory of economics imperialism still encompasses both marginalism and economization, but there is scant evi-dence to link the two. Robbins's primary contribution seems to have been to transcend the means-ends calculations of the Anglo-Irish marginalists and to have substituted for them a purely means-based analysis. However, section four demonstrates that this is not clear-cut either. The later Robbins insisted that he had intended all along to reconcile means and ends by saying that the study of hypothetical economizing agents was suitable only to the practice of economic *science*, and when economists wanted to reflect on what this meant for the world they actually lived in they had to shift register to that of political economy. Governed by such a split, the world within the model thus begins to exist increasingly in parallel to the real world.

THE MAKING OF THE ROBBINS DEFINITION

Throughout his career Robbins was unflinching in his attachment to the idea that there was only one way of doing economics, a core set of theoretical prop-ositions around which all scholarly endeavours should be organized. Susan Howson (2011: 272) has uncovered a draft appendix he eventually chose not to include in the second edition of the *Essay*, which features his very strong belief that economists were able to arrive at a consensus on what the basic objectives of the subject field are. These are propositions that seek to explore in abstract terms the ideal-typical behavioural traits which propel the equally abstract functioning of "the institutions of the Individual Exchange Economy" (Robbins 1984 [1935]: 17). For Robbins, this was economics in its entirety; other pretenders to the throne lacked legitimate claims upon it (Robbins 1984 [1935]: 114). The difficulty he identified in the early 1930s, though, was that economists were still to develop adequate ways of convincing the outside world they were practicing a unified science: "We all talk about the same things, but we have not yet agreed what it is we are talking about" (Robbins 1984 [1935]: 1). The *Essay*, then, was Robbins's call for unity of purpose and for unity in communicating that purpose. Relating general principles of theoretical progress directly to the Individual Exchange Economy was an important step in the right direction, he believed, but this first required clearing the decks of previous misconceptions. Robbins's (1984 [1935]: 1) very first comment in the main body of the *Essay* was to maintain that no progress could be made in the scientific unification of the field until the profession alighted on a "working definition of what Economics is about".

"In all my teaching and writing", Robbins (1971: 145) reflected late in his career, "this search for a coherent apparatus of analytical thought has been the underlying motive". The *Essay* must in this regard be seen as the culmination of some considerable effort, because he was ensnared by the definitional problem right from his earliest undergraduate days (Masini 2009: 428). He always had the greatest respect for the way in which the core of economic theory had advanced systematically since the marginalist revolution (Robbins 1979: xxix–xxii). Yet he had also long believed that there must be more secure epistemological foundations for capturing such a core than the early Anglo-Irish marginalists had sought in the insights of psychophysiology (Hands 2009b: 837). "[T]he unity of a science", Robbins (1984 [1935]: 2) wrote as part of his initial reflections in the *Essay*, "only shows itself in the unity of the problems it is able to solve, and such unity is not discovered until the interconnection of its explanatory principles has been established". Economics continued to betray its position as an immature science in the 1930s, according to Robbins (1984 [1935]: 22), because for all the good work undertaken by individual economists it had yet to be gathered together collectively under the banner of a convincing single problem frame.

Robbins (1936: 171) believed that Jevons had got his basic critique of classical economics correct, but that he had erred when subsequently faced with constructing something more plausible in its place. He agreed with Jevons's (1876: 619) assertion that there should be no room for theories that undermined their own precision through wilfully conflating analytical and distributional concerns; the two should be kept entirely separate. Yet he could not understand why Jevons should then insist on throwing in his hand with the hedonistic calculus, because this seemed like more of the same only in different form (Robbins 1936: 176). He made no effort to hide his contempt when writing that: "We have all heard *ad nauseum* of the deficiencies of this [the greatest happiness] principle ... How many second-rate persons and psycho-pathological major prophets have established reputations for superior sensibility by dwelling on these difficulties?" (Robbins 1964: 80). Robbins was worried that Jevons's substitution merely changed the ends to which economics was oriented but without providing adequate justification for why any discussion of ends was necessary from a purely analytical perspective. Jevons merely swapped out the objective of enhancing overall social welfare and replaced it with the objective of enhancing individual net pleasure throughout society. The revolutionary moment inherent in Robbins's work is the denial that economics had anything to do with ends at all (Robbins 1971: 147).

Robbins argued that there were merely economic means of approaching ends whose negotiation fell outside economics. In this way, his definition allows only for a discussion of the form of the argument along means-ends lines (Brown & Spencer 2012: 781). The adjudication of how shared lives might be lived well together plays no role in the analysis, being wholly displaced by a focus on how abstract techniques of choice allow individuals to conduct themselves efficiently (Endres & Donoghue 2010: 552). The focus on individual economization (the means of a means-end relationship) overrides any discussion of social objectives (the ends of a means-end relationship) (Hausman 1994: 208). Yet this relieves the core of economic theory of its genuinely economic content: economics was to become, in essence, uneconomic (Watson 2014: 9). Why, then, might such a restriction have been advocated?

Robbins's objection to how his predecessors defined the field was that it reduced economics to the mere classification of subject matter (Robbins 1984 [1935]: 16). He wished instead for economists' practice to revolve around a self-consciously analytical definition, thus moving, in Henry Phelps Brown's

(1972: 7) words, from a "field-determined" to a "discipline-determined" conception. This marked the first occasion when economics was treated knowingly as a method (see Introduction). Robbins sought to transcend the previous tendency to reserve the word "economic" to describe certain types of activity, shifting the focus instead to a single *aspect* of activity (Howson 2004: 426). He had great respect for his one-time LSE colleague, Edwin Cannan, praising him for his "weight of erudition ... [and] ... invincible common sense", saying of Cannan's younger colleagues that "we revered him" (Robbins 1971: 85, 83). Yet still he had a longstanding suspicion of what he perceived to be Cannan's overtly materialistic conception of the subject field. "[F]or the benefit of those who doubt whether they know what is meant by the term economic", Cannan (1922: 17) wrote, "I think we must fall back on 'having to do with the more material side of human happiness,' or more shortly, 'having to do with material welfare'". By contrast, Robbins insisted that economists needed to restrict themselves to asking *how* individuals make decisions when allocating resources and not *to what purpose* the decision-making process is oriented. Cannan's inversion of these priorities represented, for Robbins (1984 [1935]: 9), "the last vestiges of Physiocratic influence".

"Gradually it dawned on me", Robbins (1971: 146) wrote in his autobiography, "that the idea of material welfare was an *ignis fatuus* ... [T]he underlying fact which made so many different activities and relationships susceptible to economic analysis was the scarcity of the means with which they were concerned and not the materiality of the objectives". The word "gradually" is important in this context, because Robbins's longstanding interest in the definition of economics took time to produce a definition of its own. It took him eight years from the start of his search to change his mind over the use of a wealth-based definition and a further four before he felt ready to reveal his new choice-based definition to the world. Sensing that all is not well is not the same, it would seem, as knowing how to put it right. Robbins (1979: xii) latterly reported that he was aware from the very start of his career that there was an inherent ambiguity to the term "material welfare", but he remained prepared throughout the 1920s to call economics "the science of material satisfactions" when presenting lectures at the LSE (cited in Howson 2004: 420). "I do not pretend it is an ideal definition", he admitted to his students (cited in Howson 2011: 144), but "you are perhaps justified in saying that [economics] is the study of the general causes on which the material wellbeing or satisfaction of mankind depends". The definition that Robbins (1984 [1935]: 16) finally alighted on was very different: "Economics is the science which studies human behaviour as a relationship between ends and scarce means which have alternative uses".

However, he admitted straightaway that not all markets displayed conditions of scarcity at all times, and therefore the standard downward-sloping demand curve could only be true by convention rather than true *per se* (Backhouse & Durlauf 2009: 875). Almost certainly without knowing it, then, Robbins's definition led him to make contradictory metamathematical statements. He pointed simultaneously to a universal aspect of behaviour that could be defined into existence in aprioristic fashion and a feature of that aspect which must remain unknown until it had been empirically described. His assertion that the orthodox downward-sloping demand curve could only be true by convention located him in a world of number, responding to a Cauchian insistence that mathematical objects resemble the real phenomena to which they referred. Yet his scarcity definition also allowed for a post-Cauchian world of purely mental artefacts to be summoned as reasoning tools, where all conceivable markets could be in equilibrium together because they all obeyed the orthodox downward-sloping shape of the demand curve. The definition Robbins had laboured 12 years to perfect therefore tramples underfoot the distinction between the world within the model and the world beyond. It consequently spreads metamathematical confusion in its wake.

Robbins appeared to be unaware of these problems, remaining adamant that scarcity was still *the* most general condition available to the economist to study (Robbins 1979: xiii). The early marginalists' elevation of the individual to the centre of the subject field was retained, but here that individual is simply a chooser. They have no character, no personality and no will-governed essence, being merely a reflection of the abstract urge to economize. This meant making what you have do the best for you rather than being propelled into action by a particular goal (Robbins 1936: 209; 1979: xiv; 1984 [1935]: 16). The scope of economics thus seems to change markedly as a result of the scarcity definition, because the application of economic technique is thereafter limited only by the presence of economizing conduct, and this is by no means restricted to behaviour that takes place within obviously market settings (Spiegler 2005: 24). The domain of interest changes too, from the world beyond the model to the world within it.

There is a strong incentive, it appears, for economics to collapse in on itself under Robbins's scarcity definition, because nothing else matters than the purely hypothetical situations that emerge from pure instrumental rationality (Crespo 2013: 759). He claimed that his definition allows economists to focus on behavioural relationships, but it is questionable whether these can be thought of as genuinely human relationships, because they pay only incidental attention to how life is lived in the world beyond the model (Robbins 1984 [19345]: 17). Robbins's individuals compete in a hypothetical game that tests their calculative capabilities, but that takes them out of the broader social context which gives meaning to their existence. In a 1938 article he stated very boldly that: "I still cannot believe that it is helpful to speak as if interpersonal comparisons of utility rest upon scientific foundations" (Robbins 1938: 640). The economic agents who have economizing imposed upon them as a universal character trait have no option but to choose an efficiency strategy that amounts to minimizing their costs when dispensing scarce resources. What implications this might have for other people never enters their calculations.

Robbins's emphasis on individual choice under conditions of scarcity looks as though it formalizes the general marginalist shift towards focusing on consumption rather than on production and distribution. Yet there is a need to ask how closely these consumers resemble people engaged in actual everyday activities. It is no surprise if there turns out to be no obvious correspondence at all, because the whole idea of studying people going about "the ordinary business of life" is what informed Alfred Marshall's (2013 [1890/1920]: 1) definition of the subject field in his *Principles of Economics*. Despite latterly developing greater sympathy for Marshall's work (see Robbins 2000: 307), at the time of writing the *Essay* Robbins tended to treat him as the Cambridge Cannan. Both were susceptible to the same critique of having a classificatory definition of economics that prioritized non-scientific considerations of material welfare. Marshall allowed his economic agents to display all the behavioural motivations that impose themselves with a noticeable degree of regularity. The economist, he wrote, should "deal with man as he is: not with an abstract or 'economic' man; but a man of flesh and blood" (Marshall 2013 [1890/1920]: 22). Robbins, by contrast, sought more precision in conduct and, by settling on an economic conception of the individual as the bearer of purely instrumental rationality, allowed only the single motive of economizing into the discussion (Crespo 2013: 764). He believed he had found in the efficient disposal of scarce means the unifying feature inherent in all strictly economic endeavours (Robbins 1984 [1935]: 15). What he actually achieved was to get his professional peers to buy into disciplinary limitations in which the scarcity definition was taken as given and scholarly activity was reduced to working out the logical implications of beginning there and not somewhere else (Backhouse & Medema 2009a: 493). The pluralism inherent in Jevons's and Leslie's agreement that contrasting starting points enriched economists' practice (see Chapter 4) was unceremoniously cast aside.

The scarcity definition appears to work most effectively in bolstering economists' faith in their subject field's theoretical core. The representational relationship in operation here is not of the conventional dyad of model and target. Rather, the model is designed to reflect back to its users the status of generally accepted theoretical propositions (see Chapter 3). It is notable in this regard that Robbins (1984 [1935]: 83) drew a direct analogy between pure mechanics and pure economics. "In pure Mechanics", he wrote, "we explore the implication of the existence of certain given properties of bodies. In pure Economics we examine the implication of the existence of scarce means with alternative uses". This analogy was deliberate, because the physicist has no capacity to act upon the world when observing relationships within nature. Despite the rather flimsy ontological foundations of such a claim, Robbins (1971: 150) was adamant that he had identified in scarcity conditions relationships that also existed within nature. A step increase in his subject field's scientific credentials was therefore deemed possible, with associated increases in rigour, robustness and precision, alongside the associated exclusion of value judgements. The underlying problem with the material welfare definition, according to Robbins (1984 [1935]: 64), was that it was impossible to ask how society produced and distributed its provisioning means without also wishing to have a positive influence on life chances by acting upon the provisioning structure. This, though, equated to prioritizing social policy over economics.

THE ROUTE FROM THE *METHODENSTREIT* TO THE *ESSAY*

All definitions have consequences, but the Robbins definition appears to have had more than most. This is true for both the history of economic theory and the prehistory of economics imperialism being told here. It suggests that economists first have to interpret social situations in a particular way if they are to be amenable to being studied using acceptable techniques, as well as that the characteristics economic theory forces upon the agent matter more to the subsequent analysis than the agent's own understanding of their behavioural context. By saying that all manner of social situations can be recast as choice problems bounded by scarcity, economists have claimed the right to express social situations as economic situations before then arguing that, as economic situations, their essence can only be brought to the surface through the use of their own chosen methodology. This is all rather circular. Moreover, it entails considerable corruption of an active theory of agency, because everyone is required to follow Robbins's logical rules without regard to how they understand their own place in the world. So much, then, for Jevons's contortions over the free will question; here, it disappears entirely.

The pattern of referencing in the *Essay* offers only limited clues about how he came to settle on this position. The first edition of 1932 was more extensively footnoted than the second edition of 1935, and those footnotes suggest a heavy Austrian influence (Caldwell 1982: 103–106). Robbins was particularly close to Ludwig von Mises and Friedrich Hayek, and he seems to have used the last-minute addition of the footnotes to signal his admiration for their work. After being asked to comment on a formative draft, Hugh Dalton, Robbins's LSE colleague in the 1930s, observed that it contained "the usual superlative bouquets to Mises" (cited in Howson 2004: 441). Hayek's depiction of Menger as the hands-down winner of the *Methodenstreit* is also plainly visible (Screpanti & Zamagni 2005: 292). Despite many of the original Austrian references being deleted for the second edition, Terence Hutchison (1938: 59) set the standard of the early historiographical work on the *Essay* by arguing that the general tenor of the text remained loyal to that of the first edition. Still today it tends to be read through the perspective of Robbins returning to the scene of Menger's long-ago victories over the German historicists (Ross 2005: 90).

Howson's more recent archival work, however, has uncovered evidence of a summer school course taught by Robbins in 1929, which served as a trial run for the arguments later to be made famous by the *Essay*. The surviving lecture drafts contain almost no trace of obviously Austrian themes (Howson 2004: 427). This makes the footnotes to the first edition look like the odd one out, more the expression of a temporary interlude in Robbins's thinking than evidence of a settled opinion he carried throughout his career. Yet where does this leave Menger amongst Robbins's influences? The reverential tone might not be as pronounced as when Robbins hailed his "two brilliant disciples, Böhm-Bawerk and Wieser" (Robbins 2000: 269), but Menger remained for him the only marginalist pioneer to have brooked no compromise with the essentially theological premises of the classical economists (Robbins 1978: 23). He is also praised for having elevated to centre stage "the activity of economising", thereby providing "the *modern* origin of the conception of economics as concerned with that aspect of behaviour" (Robbins 2000: 272, emphasis in original). Nonetheless, Menger's work remains almost completely invisible in Robbins's text in any explicit sense. But maybe Robbins was never in a position to claim too close a connection anyway. At no stage in the *Grundsätze der Volkswirtschaftslehre* (*Principles of Economics*) did Menger suggest that his admittedly hypothetical individuals could be patterned into a personality type on the grounds of a single behavioural aspect. Menger wanted his market models to resemble relationships that were in evidence in the real world, Robbins apparently only in the credible-worlds constructions of economic theory.

Referential truth claims remained too important for Menger to have preempted Robbins's subsequent embrace of a one-dimensional economizing subject (Campagnolo 2010: 312). Menger and Robbins were certainly lined up in shared opposition to Jevons's position that economic behaviour was best understood on the back of the systematic collection of data on how people – or, at least, Adolphe Quetelet's *l'homme moyen* – conducted themselves in situations of choice. Both believed that the use of fundamental microeconomic principles was more instructive for the development of core economic theory than observations of agential practice (Sugden 2009b: 859). However, Menger's individuals were beset by concerted self-doubt and were still too involved in the search for their own identity to respond to any choice situation in a purely reflex manner. Jevons's use of statistical objects as reasoning tools resulted from his wish to bypass nineteenth-century psychophysiologists' fascination with the idea of autonomized habits of mind (see Chapter 4). Robbins reinstated behaviour abstracted from conscious thought into his economic theory, but not by following Menger's lead. Menger's economic agents lacked the basic self-assurance to proceed unproblematically in a pre-scripted manner. They could not populate Robbins's market models because they were ontologically incompatible with them (Grimmer-Solem & Romani 1999: 343).

In any case, when thinking about who should be credited for leading him to the economizing subject, Robbins plumped for Gustav Cassel and Joseph Schumpeter (Howson 2004: 423). Their work was shaped by Walras's *Éléments d'Économie Politique Pure*, not Menger's *Grundsätze*. Yet 40 years after the fact, Robbins (1971: 105) confessed that in the 1930s he "did not then realize how much of this was derived from Walras". There are certainly suggestions of a nascent general equilibrium approach in Robbins's work, most obviously in his idea that the presence of instinctive economizers throughout the economy leads necessarily to coordinated outcomes consistent with the reproduction of market institutions (Giocoli 2003: 124). But even in later life, Robbins (2000: 297) continued to insist that the major advances in microeconomic theory were to be discovered elsewhere. Whereas the *Éléments* presented general equilibrium as an end-state of successful market coordination, the general equilibrium mood music in Robbins is suggestive only of Austrian-style trial-and-error market processes (Blaug 1990: 185). Walras, it should be noted, is directly referenced as an explicit source of intellectual inspiration no more frequently in the *Essay* than Menger.

Vilfredo Pareto was Walras's hand-picked successor to his chair at Lausanne in 1893. He was sceptical of utility as a concept, leading him to replace Jevons's hedonistic calculus with a self-styled science of choice (Pareto 1999 [1900]: 255). For Robbins, the only thing that mattered about choice problems was that the act of choosing took place along distinctly economizing lines; Pareto had announced this as the ultimate destination for economics three decades previously. He believed that the future would belong to those who placed economic theory on a behaviouristic footing that removed the need for psychological explanations. The act of choosing was itself confirmation that a preference lay behind the choice, and no knowledge of the psychology that drove the economic activity was necessary to allow that activity to be studied (Coats 1976: 52). Simply tracing the pattern of an individual's purchases on the market would be enough to provide clues about their index of preferences, and the existence of such an index was, for Pareto, all that had to be inferred to know that the type of action in which economists are interested had just taken place (Rima 2012: 370).

Pareto's science of choice perhaps looks like a more convincing foundational text for Robbins's *Essay*, with its insistence that economists restrict their theoretical activities to studying strictly logical action (Powers 2012: 49). For Pareto, logical action should not be confused with how people make sense of the everyday situations through which they develop their own particular economic routine. It is therefore conduct that might never rise above the level of the hypothetical. By isolating a sphere in which an instrumental rationality was the only permissible precursor to action, Pareto established the conditions under which the equilibrium solutions presented by pure economic theory could pass more-or-less descriptively for the concrete phenomena with which they were most closely associated (Cirillo 1979: 32). Hypothetical economic agents were able to employ visualization techniques to reach the equilibrium position in which they could be said, from Robbins's later perspective, to be choosing the economizing course of action (Gross & Tarascio 1998: 186). Pareto latterly lost faith in this way of thinking (see Chapter 6), but what caught Robbins's eye is not where he ended up so much as his earlier argument that economic theory could be reduced to the successful adaptation of means to ends. Robbins (1984 [1935]: 23–31) pushed this argument as far as it would go by suggesting that ends could ultimately be forced out of economic theory altogether, because to be genuinely acting economically focused all attention on means.

Logical action is thus defined in Pareto's terms as the appropriation of the most efficient means to secure a given end, which in turn is defined as the equalization of the subjective and objective purposes of the individual's behaviour (Backhaus & Maks 2006: 5). It is the subjective purpose that provides the economic agent with an understanding of ends, the objective purpose an understanding of means. Robbins's (1984 [1935]: 16) criticism of material welfare definitions was that they allowed subjective reasoning to impose itself where it had no right to be. But Pareto had already made a decisive move towards banishing subjective reasoning from economic theory by bringing subjective and objective purposes into alignment on the terms dictated by the objective. Emphasizing the means of the means-end relationship hardly mattered less for Pareto than it did for Robbins. The adaptation that the individual undertakes equates to learning how to express their desires in a manner whereby a stark economic rationality out-trumps the imprints of prior socialization processes (Samuels 2012: 31).

Robbins remained fixated on the demarcation question of how an autonomous economics might be constructed, but Pareto's interest in the same question clearly waned. Over the course of his career, he found ever greater reason
to doubt the usefulness of the *Homo economicus* construction that inhabits the
domain of purely logical decision-making. In the *Cours d'Économie Politique*
(*Course in Political Economy*), published in two volumes in 1896 and 1897, the
descriptive capacity of the *Homo economicus* construction is largely taken as
given, but in the *Manuale di Economia Politica* (*Manual of Political Economy*)
of 1906 it is clear that its core behavioural features were to be denied a place
in descriptions of the world beyond the model. The more Pareto came to
accept that human action must itself be synthetic in nature to reflect the multiple dimensions of identity (Tarascio 1972: 421), the more he appears to have
doubted the efficacy of abstracting out of economic theory all non-logical
features of basic experience (Pollini 2001: 34). Hence there is evidence in his
work when it is read holistically of increasing scepticism that the social world
divided neatly to allow each discipline to generate strictly compartmentalized
knowledge (Wood & McClure 1999: 9). He latterly described as an "absurd-
ity" the proposition that economic theory could explain everything of interest in economic action, let alone everything *per se* (Pareto, cited in McClure
2002: 48). Robbins's unflinching commitment to demarcating economics as an
autonomous subject field dedicated to studying a separate category of human
behaviour seems to be a good example of the bogus scientific unification the
later Pareto warned against.

WICKSTEED'S INFLUENCE ON ROBBINS'S *ESSAY*

Robbins's position relative to both the Austrian and Lausanne Schools ultimately seems to be mediated by the influence of the second generation English
marginalist, Philip Wicksteed. The eulogies to Wicksteed – "a very considerable figure" (Robbins 2000: 283) capable of writing "a masterpiece of systematic exposition" (Robbins 1933: 200) – go beyond anything he said about
anyone else from the previous generation. Indeed, Robbins appears to have
come only comparatively lately to a thoroughgoing study of the early marginalists, familiarizing himself with the content of their work only after having

learnt about it second-hand through the thoughts of his direct predecessors (Robbins 1971: 105–08). It is perhaps only natural that someone who held such a linear conception of his subject field would want to begin with what he took to be the most convincing contemporary viewpoints. This is what drew him to Wicksteed's 1910 book, *The Common Sense of Political Economy*, after which reading the early marginalists merely confirmed the good judgement he had already identified there (O'Brien 1990: 163).

The intellectual historian Terence Hutchison (1953: 95) describes Wicksteed as Jevons's "one great disciple", with Wicksteed (2014 [1888]: vii) confirming that *The Theory of Political Economy* was his "guiding principle". Wicksteed's stated aim was to take Jevons's work in a direction that placed increasing emphasis on mathematical content to make his economic arguments (Flatau 2004: 74). Robbins's position on the mathematics issue was something of a moving target. He remained adamant that Marshall overplayed his hand in advocating the use of ordinary language in economics, and that he therefore deserved criticism for failing to place his own theory on a much more obviously formal basis consistent with enhancing the clarity of the subject field's core (Robbins 2000: 307). However, he became increasingly sympathetic to Marshall's (2013 [1890/1920]: 85) argument that mathematics, used judiciously, might help to explain economic principles but should not be confused for those principles themselves. "Now, in my extreme old age", he told his students, "I agree with Marshall, and in consequence I dare say that my name will be mud with all sorts of much more eminent contemporaries than I pretend to be" (Robbins 2000: 308). The early Robbins seems to have accepted without hesitation the need for economic theory to be underpinned by mathematical *logic*, but he was wary of too much mathematical *content*. The later Robbins took his concerns about mathematical content to a degree where it was no longer obvious that he still consented to the use of mathematical logic at all. So why was he so sympathetic to Wicksteed, a self-trained but nonetheless highly accomplished mathematician within the economics of his day?

It is not the mathematics *per se* that seems to have caught Robbins's eye, so much as what Wicksteed was able to get the mathematics to *do*. Wicksteed had taught himself the techniques of differential calculus in an attempt to get right to the heart of Jevons's analysis (see Wicksteed 2014 [1888]: 39). He did so in a way that echoes Augustin-Louis Cauchy's (2009 [1821]: 21–48) pioneering efforts to impose definitional clarity on the calculus (see Chapter 1).

Cauchy stopped well short of succumbing to the Lagrangian temptation to liberate mathematical activities from substantive meaning linked to concrete realities. Wicksteed did likewise, but his preference for ordinal scales of measurement was nonetheless changing very profoundly what numbers signified in mathematical market models. By the end of Wicksteed's career techniques in calculus were significantly more advanced than they had been in Jevons's day, providing him with insights derived from mathematical logic that were simply unavailable to Jevons (Creedy 1998: 128). Yet the lines between purely abstract and now obviously concrete applications of differential calculus were becoming increasingly blurred. For his part, Wicksteed seemed to want to use the developments both to add greater precision to the marginal concept and to apply it more extensively. The mathematical exposition was something of a distraction for the older Robbins in particular, but maybe it was a necessary distraction if it was to take economics further in his preferred direction.

Even here, however, there are important historical discontinuities. Important changes were incorporated into Wicksteed's thinking in the 22 years between the publication of *The Alphabet of Economic Science* (1888) and *The Common Sense of Political Economy* (1910). The early Wicksteed saw his task to be the fuller explication of Jevons's system, but Robbins embraced the later Wicksteed precisely because of the deliberate distance he opened up with Jevons's *Theory of Political Economy*. In a revealing passage from the *Essay*, Robbins lamented the "incautious utterances" of "certain of the founders of the modern subjective theory of value", having claimed "the authority of the doctrines of psychological hedonism as sanctions for their propositions ... [T]he names of Gossen, Jevons and Edgeworth, to say nothing of their English followers, are a sufficient reminder of a line of really competent economists who did make pretensions of this sort" (Robbins 1984 [1935]: 84). The early but not the later Wicksteed was undoubtedly one such "English follower".

Jevons (2013 [1871/1879]: 58) deliberately circumscribed the scope of his economic method by proposing an essential difference between what he called the higher and lower motives. The higher were associated with a life lived well, the lower with maximizing pleasure relative to pain. He used his *Principles of Science* (Jevons 2010 [1887]) to urge audacity in the application of mathematical methods beyond their usual limits, but he consistently argued throughout *The Theory of Political Economy* that importing mathematically tractable economic techniques to study the higher motives was a step too far.

Philosophical concepts that helped to set the context for the good life – justice, beneficence, tolerance, trust, respect – were therefore to remain outside economics. It was not necessarily the principle of forcing ethical decision-making into a series of differential equations that bothered him, so much as the fear that ethical norms were insufficiently well understood in substantive terms to give full licence to the available mathematics (Spiegler 2005: 3). Current understanding in Jevons's day suggested that ethical behaviour was simply too distinct from economic behaviour to warrant scientific unification using the same explanatory models (Daston 1978: 196).

The early Wicksteed (2014 [1888]: 46–53) had pushed beyond Jevons by declaring marginal utility to be the characteristic in utility on which human choices ultimately depend. However, both thought that utility-oriented behaviour could be captured strictly as a quantity (Udehn 2001: 53). Measurement techniques were insufficient for putting real numbers reflecting real feelings on increases in marginal utility, but the utility concept had to be thought about quantitatively nonetheless. A concrete conception of number – understood in a cardinal sense as per the real number line – was still to be the mainstay of mathematical economics at this stage of Wicksteed's career. It was a similar concern for concrete measurability that had encouraged Jevons to restrict economics to only the lower motives governing human existence. The simplest numerical measures of utility were much easier to accept as plausible accounts of want-satisfaction in relation to the demand for goods than in relation to support for particular conceptions of the good life (Schabas 1990: 127). By the time of his *Common Sense*, however, the later Wicksteed's marginalist emphasis had changed from trying to calculate the amount of utility that was available at the margin to trying to understand merely how choices might be said to be made there.

As Robbins (2000: 278) noted, Wicksteed had been the first English-speaking economist to develop a fully thought through theory of opportunity cost, with Menger's student Friedrich von Wieser being the first overall. In the *Common Sense* this was presented as the alternative cost doctrine, where each decision could be considered to have a price in monetary terms but also a price in terms of foregone possibilities (Flatau 2004: 95). Every decision the individual might be called on to make could be placed somewhere on the spectrum covered by this dual sense of price. It was a complementary idea about shadow prices, it should be recalled, that triggered Paul Samuelson's initial use of Lagrange multipliers to bypass the need for empirical treatment of real phenomena (see Chapter 1). Marginalism could therefore be described not only as a universal principle of human nature but also one where there was no need to distinguish between what classificatory definitions of the subject field had treated as different economic and non-economic motives for action (Wicksteed 2003 [1910]: 194, 431). There are important implications for the much later turn towards economics imperialism in Wicksteed's efforts to divert marginalist mathematics from a cardinal to an ordinal understanding of number.

Robbins particularly approved of how the later Wicksteed had been able to remove all allusions to the ability to quantify utility and instead wrote about preference orderings. He identified this as the moment when the outline of economizing agents properly began to take shape in his imagination (Robbins 1971: 146, 1979: xxi). The early Wicksteed's (2014 [1888]: 81) focus on the calculable "differential coefficient of the total utility" was replaced by the later Wicksteed's (2003 [1910]: 36) merely formal emphasis on the "scale of preferences". "We are not obliged to be constantly considering alternatives, because in a fairly well regulated mind the suggestion of any particular item of expenditure does not as a rule arise until it is approximately in its proper turn and place for gratification. The vague sense of restraint, which subdues and suppresses it, is really the unanalysed consciousness of the higher place on the scale of preferences of certain other unspecified items which will one by one assert themselves in due time and place" (Wicksteed 2003 [1910]: 35–6). Robbins (1933: 206, emphases in original) neatly summed up the difference in the following way: "The *Alphabet* starts from the idea of the rate at which total utility is increasing; the *Common Sense* from the positions on the relative scale of preferences which marginal units of different commodities occupy. In the *Alphabet*, in spite of the earlier recognition of the relativity of the utility concept, utility is treated as if it were something absolute and measurable. In the *Common Sense*, the sole relevance of *relative* utility is emphasized and the idea of *measurability* tends to give place to the idea of *order*".

The early Wicksteed's (2014 [1888]: 54) stress on the potentially calculable "hedonistic value" (to which Robbins objected when it was suggested that this underscored every aspect of the individual's demand function) thus ceded priority to a more formal and consistently subjective approach to the choice puzzles faced by that individual (to which Robbins strongly assented) (Wicksteed 2003 [1910]: 27). This led to a much more expansive conception of the scope of economic analysis, whereby the general principle of selection between alternatives put paid to the notion of a distinctly economic realm (Wicksteed 2003 [1910]: 3). "[W]hatever our definition of Economics and the economic life may be", wrote Wicksteed (2003 [1910]: 160) in the *Common Sense*, "the laws which they exhibit and obey are not peculiar to themselves, but are laws of life in the widest extent". Furthermore, "all the heterogeneous impulses and objects of desire or aversion which appeal to any individual, whether material or spiritual, personal or communal, present or future, actual or ideal, may all be regarded as comparable with each other" (Wicksteed 2003 [1910]: 32). This looks as though it could act as the calling card for all contemporary economics imperialists, even if none of them is remotely familiar with Wicksteed as a historical bridge to how they think today.

One way of enforcing such comparability is to impose upon every individual the characteristics of an abstract *Homo economicus*, but this was not the route taken by Wicksteed. Another discontinuity emerges in the prehistory of economics imperialism, because he rejected Pareto's use of given indifference curves and consequently forwarded a much more complex configuration for human agency (Drakopoulos 2011: 461). The later Wicksteed's economics promoted realism through practical reason (Comim 2004: 479), which was a point that seems to have rather escaped Robbins. He warned his students: "Don't, for goodness sake, be put off by the title of Wicksteed's book", stressing that what the reader finds within the *Common Sense* is sophisticated elaboration of complex theoretical principles built on the opportunity cost doctrine, not something of folksy intuition as implied by the title (Robbins 2000: 278). Yet Wicksteed (2014 [1888]: 48) managed to combine the two, using homespun examples drawn from the everyday life of a materfamilias to illustrate his conviction that in all sorts of circumstances people do, whether consciously or not, act as if they are seeking to equalize opportunity costs on the margin.

Wicksteed (2003 [1910]: 87–91) appealed to the image of the conscientious Victorian housewife he remembered from his youth, inviting his reader to picture her serving meals in deliberately unequal proportions as an instinctive calculation of how to ensure that everyone – including the family cat – has enough to eat. His housewife's homemaking skills appear intermittently, each time the reader has to be reminded that even the most apparently mundane tasks of resource administration are subjected to marginalist principles. She is also placed in wider contexts, whereby the objects of choice that shape the everyday experience of family life are extended to include concern for distant strangers as well as for one's immediate family, one's social position in the eyes of others, and the respect that is due both to the self and to one's acquaintances (Steedman 1989: 107). Even the most seemingly dissimilar choice situations can be compared at the margin, according to Wicksteed (2003 [1910]: 368), as long as the economic agent has the Marshallian characteristics of "flesh and blood" (see Marshall 2013 [1890/1920]: 22).

Robbins, however, was not prepared to follow everything the later Wicksteed had written. He displayed no ambition to mimic Wicksteed by inserting a recognisable level of realism into his economic agents, worried maybe that this concern for individuals in complex social situations is what Marshall (2013 [1890/1920]: 79) and Cannan (1922: 17) used in continuing to keep economics anchored to a classificatory definition. "Perhaps", writes Robert Sugden (2009b: 864), "Robbins is embarrassed to find concrete facts of experience, such as how a housewife makes puddings and feeds the cat, intruding into the abstract spaces of pure economics". Whatever the reason, Robbins's economic agents were selected for how well they complemented his alternative analytical definition that revolved around scarcity constraints, but they were much less fully human as a consequence (Wilson & Dixon 2012: 11). However, this conception of an abstract individual did allow him to channel other aspects of Wicksteed's work in stressing the potentially limitless applicability of its underlying frame of reference.

POLITICAL ECONOMY VERSUS ECONOMIC SCIENCE IN ROBBINS

Given the rather strident tones in which the *Essay* was written, it is noticeable that reticence remains the abiding feature of Robbins's claim to success. "I assure [the doubters]", he admitted in the Preface to the second edition, "I am not at all cocksure about any of my own ideas" (Robbins 1984 [1935]: xxxiv). Even half a century after he had first trialled the outline of his definition, and three decades after it had become the settled professional opinion of what economics was about, he felt compelled to return to the same theme. Perhaps it will be possible, he told an assembled audience gathered in his honour in his eighty-first year, "to put things in such a way as to make peace with some of my critics" (Robbins 1979: xi). His early reluctance to declare victory for himself was almost certainly justified in the midst of what Bob Coats (2014: 76) calls the "storm of criticism" that the *Essay* provoked. He was variously accused of excessive abstraction from the everyday economic lives that anyone might possibly be expected to lead, excessive rigidity in the behavioural motives that economic agents were allowed to display, and excessive formalism of method that led to an equally excessive narrowness of what economists might legitimately say. This is quite some charge sheet.

Following the publication of the *Essay*'s first edition in 1932, his contemporary critics accused him of having advocated a purely formalist economics. Barbara Wootton (1938: 61) led a second wave of similar concerns following the publication of the revised edition of 1935, arguing that the more economists were encouraged to emphasize the form over the content of their arguments, the less rooted in reality their theories would be. Terence Hutchison (1938: 54) weighed in, saying that to replace the previously richly textured tradition of studying concrete economic situations with a single focus on the implications of economizing behaviour was tantamount to relieving the field of "all facts". Robbins's scarcity definition, he had previously contended, confused a tautology (*if* scarcity is a contextual factor, *then* maximizing agents must economize) for an inference (*since* people are usually subjected to scarcity constraints, *therefore* they have an incentive to economize). Inferences alone are capable of connecting economic theory to economic reality, but "this requires some data to be known in order to establish a 'since'" (Hutchison 1935: 161). The subsequent philosophy of science literature was obviously unavailable to Robbins's critics in the 1930s. Yet they were strongly hinting that he was transposing judgements about the acceptability of economic theory from how it performed relative to the world beyond market models to how it performed relative to the world within those models. The inferential capacity of market models was also refocused from revealing more about how actual lives were being led to revealing more about the precision of theoretical claims.

The critics objected to the apparently Misean emphasis on an unbridgeable gap between positive and normative economics (Spengler 1934: 316), believing that Robbins's focus on neutrality between ends left them with almost nothing purposive they could say *as* economists (Cannan 1932: 425). The ability to speak on social matters, the critics argued, was more than just a supplement to economists' day job of refining fundamental microeconomic principles (Fraser 1932: 569). Rather, it was a constitutive aspect of what drives the theoretical endeavours in the first place (see also Hawtrey 1926: 61; Harrod 1938: 396). It should be remembered (see Chapter 4) that it was on precisely this issue that so many of the marginalist pioneers rebelled against their inheritance, believing that the economics of the late nineteenth century had failed the social goal of improving the lot of the poorest sections of society, often writing as a consequence in terms of their predecessors' betrayal of the public (Winch 2002: 5). The Jevons–Wicksteed–Robbins lineage becomes somewhat muddied in this regard, seeing as Jevons (1876: 619) believed that economists got what was coming to them when they were "regarded as cold-blooded beings, devoid of the ordinary feelings of humanity". There was clearly to be no neutrality between ends in his conception of the profession.

Yet economic theory, Robbins (1984 [1935]: 39) wrote, consisted only of the generalization of the "inevitable relationships" that arise whenever anyone is confronted by scarcity. From this perspective, fundamental microeconomic propositions deal with existence theorems, and they are consequently very different to wanting to know how best to improve overall levels of human welfare (Robbins 1979: xviii). The *Essay's* "'extreme neutrality' thesis" (Davis 2005: 192) provides ample instances in which economists are being encouraged to purge their work of its ethical content (Backhouse 2009: 477). This, at any rate, was how it seemed to the *Essay's* early reviewers, who poured scorn on its reduction of economics to technical adaptations to scarcity constraints (Dobb 1933: 590). Lindley Fraser (1932: 557) worried that it turned rationality into an end in itself, undermining Robbins's objective of removing from economic theory all discussion of ends. Ralph Souter (1933a: 387) complained that economists would be "industriously hamstringing themselves" were they to follow Robbins's suggestion to place logical precision above their ability to enact good in the world. It seems as though any discussion of the ethics of competing economic norms had to be rehomed elsewhere. Robbins took aim at Pigovian welfare economics, with its focus on changing the social conditions under which positions of Pareto optimality could arise, so that net utility gains might be possible for society as a whole (Pigou 1912, 1920). He was no more enamoured of the work of Ralph Hawtrey and John Hobson (Robbins 1927: 176), who were chastised for wishing to study the economy in relation to the ethical systems that legitimated its institutionalized customs and norms

(Macciò 2015: 177). None of this was legitimate terrain for economic theory, according to the scarcity definition.

The mature Robbins, however, presented a more nuanced depiction of the field, whereby extreme value-neutrality might still be allowed to coexist with ethical position-taking. Indeed, he was adamant that there was always more potential for moving between registers than the *Essay*'s critics permitted themselves to see. "All that I had intended to do", he suggested in his autobiography, "was to make it clear that statements about the way in which an economic system worked or *could* work did not in themselves carry any presumption that that was the way in which it *should* work" (Robbins 1971: 147, emphases in original). He then forwarded an account of the "non-overlapping domains" of economic science and political economy (Scarantino 2009: 464). It was only necessary to remove value judgements from the former, Robbins argued.

A rather convoluted position thus emerges. Hawtrey, Hobson and the Cambridge welfare economists following Pigou were all in error, it seems, for saying that ethics should be considered an integral part of how economists conducted their theoretical reflections. At the same time, they ran the risk of muting their own voice in public debates if they paid no attention to the cultural norms through which everyday economic life is produced. Economists must engage ethics, in other words, but not in the name of enabling economic theory to progress. Economists should have ideas about the world they want to see being brought into existence beyond their market models, and they should be prepared to defend those ideas in front of the public, but they do not belong in the province denoted by the term "economic science": "I went out of my way to say that … it is only if one knows how the machine runs or can run that one is entitled to say how it ought to run" (Robbins 1971: 148). Once again, though, we see ontological oscillation about the purpose of economic theory. Even if it is clearly not about enacting visions that operate beyond the Individual Exchange Economy (how the economy "ought to run"), there is an evident slippage between asking it to model lived experiences (how the economy "runs") and abstract propositions (how it "can run"). Inferences to the real world via economic theory and inferences to a credible world constructed *by* economic theory are not the same thing.

Already in the Preface to the second edition of the *Essay* in 1935 Robbins felt it necessary to respond to the charge that he was telling economists to desist from activating their consciences. According to Fabio Masini (2010: 43, 2009:

421), no criticism of the first edition hurt him more than that his preferred "absolutist approach" to the boundary separating economics from ethics prevented economists from holding opinions about the state of society. Robbins's response was to argue as clearly as he could that economists who possessed wider knowledge were of considerably more use as members of society than those who did not. "[B]y itself", he argued, "Economics affords no solution to any of the important problems of life" (Robbins 1984 [1935]: xxxvi).

Another boundary thereby comes into view, but this time within economics. If Robbins was seeking to remove value judgements merely from economic *science*, then when the embrace of ethics associated with political economy is added to the mix, the intellectually rounded economist to whom he pays such respect seems much more likely to emerge. This is presumably what he had in mind rather than the literal wording of the text when writing that "an education which consists of Economics alone is a very imperfect education" and that "an economist who is only an economist … is a pretty poor fish" (Robbins 1984 [1935]: xxxvi). It is also what he meant when discussing the role of economic theory with his students: "I have urged you in the interests of clarity of thought to keep its conclusions free from judgements of value … But I have not urged you to *ignore* such considerations. I have not recommend[ed] you to be merely economists" (Robbins, cited in Howson 2004: 432, emphases in original). Economic science was where academic reputations were to be built, but political economy provided an important reminder that those reputations were of limited worth if the knowledge on which they were founded lacked traction on practical matters. The economist could therefore remain trapped within practices requiring merely the elucidation of logical implications when acting in their immediate professional environs, but they should quickly cast off "the cloak of economic science" whenever the public interest demanded it (Balisciano & Medema 1999: 260).

It is important to note here, though, how the relationship between economic science and political economy is deemed to work. The latter function exists only insofar as it first passes through the former. It is impossible to imagine in Robbins's construction a defensible political economy position that is not first founded on reputable economic science studies of fundamental microeconomic propositions. Everything eventually comes back to choice under conditions of scarcity. Economic science is so much the more important of the pair that under Robbins's schema political economy cannot even be called into existence without it (Freeman, Chick & Kayatekin 2014: 519). All possible outlooks on the world must therefore first be routed through economic science, making John Neville Keynes's attempts to resolve the Anglo-Irish *Methodenstreit* a more likely influence than Jevons's on how Robbins viewed economists' internal division of labour (see Chapter 4).

This same sense of hierarchy also features prominently in Robbins's account of the relationship between economics and its fellow social sciences. Attention to the world within market models also takes priority here over attention to the world beyond. The whole rationale for engaging in economic science, to his mind, is to perfect a style of analysis that renders economics distinct and, on matters of analytical precision in particular, superior to all other social sciences. Any attempt to marry economics with anything else is therefore a retrograde step diluting its essence (Colander 2009: 443). In rather forthright terms late in his career, Robbins was still dismissing as "nonsense … sterile chatter" the argument for Souter-style disciplinary cross-fertilization (see Chapter 1). This argument, he scoffed, presumably in the direction of the *Essay*'s early critics, belonged only to "those who have no sense of direction of their own" (Robbins 1971: 89).

It is this attitude that Souter (1933a: 386) had immediately warned would lead to a "fastidious withdrawal from organic intercourse" between the social sciences. It promoted only a "pathological mutual distrust" amongst those who aligned themselves to different social science silos (Endres & Donoghue 2010: 553). Robbins was more than happy to annex for economics the study of the economizing behaviour that is the rational response to scarcity constraints, while declaring that economists could have nothing to say in their scientific endeavours about the social processes through which economizers' ends were established. The determination of preferences was therefore placed beyond the boundaries of economics and, from the perspective of economic theorists, into a black box. To make sense of the social processes through which demand-side conduct is enacted, economists post-Robbins were required to take other people at their word for what was going on, or ignore the issue altogether. They had no independent theory of preferences of their own to call upon (Backhouse & Durlauf 2009: 875).

It was never likely that Robbins would backtrack on the question of demarcation under pressure from his critics. After all, he had confessed in a letter written in the 1950s to the American economist, John Maurice Clark, that the ultimate rationale of the *Essay* was as "a sort of preliminary manifesto designed to forestall the criticism that I did not know where the borderline between the different disciplines really lay" (cited in Howson 2004: 417). He certainly stuck to his guns following criticism of the first edition by continuing to recommend "the abstention of the economist from all interest or activity outside his own subject" (Robbins 1984 [1935]: xxxv). His fascination with the demarcation problem long preceded the writing of the *Essay* (Masini 2009: 423), seemingly starting as early as 1919, when upon demobilization he attended a series of lectures organized by the Workers Educational Association (Howson 2004: 418). His attempt in the 1930s to place sociological questions definitively outside economic theory resulted from his belief that sociology was merely an alternative route for smuggling psychological premises back into economics. Robbins was perhaps at his most Austrian when refusing to countenance any backsliding on the desire to exclude psychology wholesale from economic theory. "[A]ll that is assumed in the idea of the scales of valuation", he wrote, when presenting his anti-hedonistic account, "is that different goods have different uses and that these different uses have different significances for action, such that in a given situation one use will be preferred before another and one good before another" (Robbins 1984 [1935]: 85).

Psychological categories such as preferences remained integral to Robbins's conception of economic science, but now stripped of any genuinely psychological content (Giocoli 2003: 86). Economizing agents can be said to prefer A over B, even though economic theory now had no position on what it is about A that allows such a preference to take shape. The individual will always be able to explain to themselves why their preferences are as they are, but this explanation, according to Robbins, is of no concern to economic theory. Economists, he wrote, must not be "bamboozled into believing [... that ...] fashionable psychology" assists the search for enhanced theoretical precision (Robbins 1984 [1935]: 84). Science progresses instead through "accuracy in mode of statement", whereby "one of the greatest dangers which beset the modern economist is preoccupation with the irrelevant" (Robbins 1984 [1935]: xxxi, 3). This cast Robbins himself as the chief border guard, inviting his peers to talk about issues that attracted the attention of other social scientists only as political economists, and even then it was necessary for them to do so using a distinctive view that they had learnt from economic science.

CONCLUSION

Robbins's strict demarcation of the terrain of economic science was simultaneously the narrowing of economics so that it no longer needed an account of motivation to explain the nature of economic outcomes. For him, this was far too reminiscent of the general argument pattern of the classical economists that marginalism had supposedly eclipsed. "[T]here is no doubt about Jevons's originality", Robbins told his students late in his career, "but Jevons was more of a classical economist ... than he knew" (Robbins 2000: 295). Reliance on the hedonistic calculus might have seen him offer a different basis to his explanation of economic content, but the adaptation of Benthamite utilitarianism meant that he was still arguing on the terrain of motive-laden behaviour that was so central to his classical inheritance.

Wicksteed fared better in Robbins's judgement, at least when eventually moving towards specifically ordinal conceptions of utility by replacing the hedonistic calculus with valuation scales. It was not just the case that Wicksteed had produced the "most exhaustive ... exposition of the technical and philosophical complications of the so-called marginal theory of pure economics which has appeared in any language" (Robbins 1933: 198). Here, at last, Robbins could conclude that the preceding general argument pattern had been superseded. But even Wicksteed's later work, the most compelling bridge between Jevons and Robbins, was built upon an optimistic reading of the possibility of mutually beneficial interdependence of the social sciences. This was a possibility that Robbins refused to entertain as he sought to establish an essential autonomy for economics (Steedman 1989: 117; Sugden 2009b: 863). He considered it to be no great gain if liberating economics from the teachings of nineteenth-century British psychophysiologists was only to replace it with content originating from somewhere else. Wicksteed's *Common Sense* might have identified the potential for developing a conception of motiveless agents, but his own turn to everyday homilies to illustrate economization agency in action invoked knowledge about motivation that stood outside economic theory. Scientific unification remained Robbins's ultimate goal, in the first instance the unification *of* economics, before it would be possible to start thinking about unifying other branches of knowledge under the settled opinions of economic science. He was adamant, though, that the initial task was for economists to act in isolation.

The lasting significance of Robbins's *Essay* is that it established a series of methodological rules which, if followed closely, made it impossible to study social phenomena as they were actually experienced (Hodgson 2008: 107). The scientific unification towards which economists should strive, it seems, was antithetical to preserving the essential unity of social life. Robbins's salami-slicing approach to the individual's interactions with the wider social whole simply would not permit it. As a result, once the scarcity definition became firmly established, economists found it much more difficult to refer directly to the social processes through which any functioning economy is institutionalized (Schabas 2009: 8). Robbins turned economics into the study of causal laws that take the form of rational ideal-types (Lawson 2003: 159). Studying the implications of abstract rationality is not the same, of course, as studying real people in action. The market models of the former exist on a different plane of construction to the market models of the latter.

The two most important observations ever made about Robbins's work thus remain objections raised in the immediate aftermath of the publication of the *Essay*. Souter (1933a: 379) criticized Robbins for having restricted economics to the formal working through of the repercussions of a behavioural logic unknown in any existing state of society. Hutchison (1935: 160) criticized him for having mistaken a tautological for an inferential understanding of that logic. The result has very definitely been an escape from the empirical. It was no part of Robbins's thought process when advocating such a move, but this was entirely consistent with the prior dethroning of Cauchian metamathematics. The *Essay*'s insistence that market models no longer had to incorporate empirical reality paved the way for those who had superior mathematical skills to edge towards purely defined functions. The explanation of fully described economic behaviour involved asking a series of "why" questions that would take the analysis beyond where Robbins wanted to place the boundaries of economic science, limits consistent with what the early Pareto had called the logical domain. In rejecting the empirics of everyday life, Robbins also managed to turn economics away from a concern for referential truth claims.

A qualitative change thus ensued post-Robbins in the potential for enacting scientific unification through the use of economists' market models. Prior to the publication of the *Essay*, such a process seemed destined to be, at most, something approaching a Newtonian approach in which the master model would be based on conditions drawn from experimentally derived observations of real phenomena (see Chapter 1). Each of Jevons, Menger and Walras in their own way had argued for building market models on the basis of numbers that invoked a representational relationship with everyday realities. Robbins's escape from the empirical removed the need to calculate parameter estimates for market models on the basis of observation. Following the publication of the *Essay*, the potential for scientific unification changed location from a Newtonian to a Maxwellian world. Formal proof-making could now come to the fore, because formal rather than referential truth claims were in the ascendant. Robbins's scarcity definition created the opportunity for such a shift, but Robbins was not to be the person to enact it. His distrust of mathematical explanation, coupled with his lack of high-end mathematical capabilities, meant that it was always going to be a task for someone else.

Formal axiomatization of market models in a Hilbertian sense therefore passes indirectly through Robbins's adoption of aprioristic techniques. His "naïve introspectionism" (Blaug 2009: 418) involved the embrace of situation-defining assumptions about how people would order their choices were they to exhibit the same rationality in the world beyond the model as in the world within the model (Hands 2009b: 833). This becomes crucial to the move between the second and third phases in the prehistory of economics imperialism, because it provided Paul Samuelson with a point of departure for an even more striking break with the early Anglo-Irish marginalists. However, Samuelson saw his own work as a complete disavowal of Robbins's. He treated the *Essay* as merely another statement in the Austrian tradition, whose "*a priori* truths" he described as "a delusion" (cited in Pizano 2009: 112): "I completely repudiate the opinions of thinkers such as Lionel Robbins", he stressed. Nonetheless, the prehistory of economics imperialism still seems to encompass the work of both Robbins (in its economization phase) and Samuelson (in its maximization phase). Similarly, it also seems to encompass the work of both Jevons (in its marginalism phase) and Robbins (in its economization phase), even though Robbins's *Essay* was written as a wholesale rejection of the image of economic theory that was closest to Jevons's heart. Meanwhile, the boundaries between Cauchian and post-Cauchian metamathematical worlds appear almost hopelessly indistinct in what Robbins borrowed from Jevons and Samuelson from Robbins, while denying any such inheritance. This adds further force to the claim that what today we know as economics imperialism had anything other than a smooth journey to its birth.

CHAPTER 6

Maximization as proto-imperialist move 3: Paul Samuelson and the search for operationally meaningful mathematical theorems

INTRODUCTION

The 1970 Nobel laureate in Economics, Paul Samuelson, was a strong devotee of the so-called SMMS narrative of disciplinary development, whereby he sits at the apex of a lineage that passes uninterrupted through Adam Smith, John Stuart Mill and Alfred Marshall (Feiwel 1982: 3). A series of advances are posited through which the best bits of the preceding theory were preserved in successive conciliatory syntheses, shorn of their troublesome features. "The giant steps that have been made in this narrative are steps of codifications of the body of knowledge", writes Till Düppe (2011: 41) of Samuelson's approach to the history of economic thought. "The disagreements economists had with their predecessors did not have to be argued, but simply vanished in their recodification." Samuelson was only too happy to assume the role of recodifier-in-chief, because the only justification for studying what economists once did, he said, is to understand how they were now doing the same thing better. He described his own approach through the "good, if ugly title [of] … 'Presentistic history'" (Samuelson 1991: 6), as well as through the potentially better if definitely even less appealing "Cliowhiggism" (Samuelson 1987: 56). "Inside every classical writer", he argued, "there is a modern economist trying to get born" (Samuelson 1992: 5). From this perspective, the person responsible for the last recodification must necessarily be best placed to speak on behalf of the whole subject field, both now and for everyone from the past who has helped it get to its current position.

The text that allowed Samuelson (1983a: xxvi) to claim the status of disciplinary standard-setter was his 1947 *Foundations of Economic Analysis*. Others have hardly been less stinting in their praise for it. *Foundations* has been described by luminaries within the profession as "a remarkable performance" (Fischer 1987: 236), "the most important book in economics since the war" (Kenneth Boulding, cited in Lucas 2001: 7), the text that "really formed [the successor] generation of economists" (Robert Solow, cited in Breit & Hirsch 2009: 156). The Royal Swedish Academy of Sciences, when announcing that it was awarding Samuelson the Nobel Prize, said that through this one publication he had done more than any other economist in "raising the level of analysis in economic science" (Nobel Committee 1970).

However, his anointment as disciplinary figurehead has by no means gone unchallenged. For every economist who has been willing to describe Samuelson as "the master craftsman of modern economic techniques" (Dransfield & Dransfield 2003: 104), there has been another who has said that he is merely the purveyor of "inspirational poetry [... which has ...] few practical benefits but seeks to uplift people's spirits with a vision of the power of the human mind" (Nelson 2001: 147). For every economist who has lauded "the boldness with which he advocated the use of mathematics" (Backhouse 1994: 214), there has been another who believes that the result is "composed of a host of seemingly self-contradictory propositions" (Mirowski 2002: 226). For every economist who is adamant that he is the person "who has most worked to make economics into a science" (Pizano 2009: 109), there has been another for whom he has created "truly ... the economics of nowhere" (Hodgson 1999: 37), "an entirely imaginary and fanciful economic cloud cuckoo land" (Blatt 1983: 171). These decidedly mixed reviews present fundamentally different ontological objectives for economic theory: whether its role is to produce ever more precise accounts of the world within the model or greater insights into actual life experiences. From the former perspective, Samuelson might well have provided the first glimpse of a future to which a sizeable proportion of his professional colleagues would choose to defer; from the latter, the crowd is always capable of leading itself astray. As Deirdre McCloskey (1990: 231) has suggested: "Since 1947 ... we in economics have been on a wild goose chase to find theorems provable by mathematical means that will miraculously give us a purchase on the world without having to venture out into it".

The debate about Samuelson's legacy reduces to whether advances in mathematical technique are worth the cost of forgone economic relevance. *Foundations* certainly brought a previously unwitnessed level of mathematical precision to economic theory (Hands 2019: 49), but it also seems to have elevated derivational unification in the world within the model above all other explanatory goals (see Chapter 3). For a century from the 1840s, cutting-edge economic theorists had shown a propensity for *thinking* in mathematical terms, but they had seemed to recoil for ontological reasons from translating this directly into *arguing* in mathematical form. Samuelson's focus on the proto-imperialist move of maximization – the third phase of my prehistory of economics imperialism – transcended such a split. He announced in the Introduction to *Foundations* that there could be no return to the "laborious literary working over of essentially simple mathematical concepts such as is characteristic of much of modern economic theory" (Samuelson 1983 [1947]: 6). Mathematical thinking and mathematical argumentation were henceforth to be conjoined. Right from his earliest writings he had wanted to put his predecessors on trial for not having had the courage of their mathematical convictions. It was not without reason, then, that Joseph Schumpeter turned to his fellow examiners having listened to Samuelson defend his Harvard PhD thesis in 1941 and is reputed to have asked, "Well ... have *we* passed?" (cited in Nasar 2011: 418, emphases in original). Why be imprecise in the formulation of economic problems, Samuelson asked, when other disciplines had already shown the way in harnessing mathematics to unify the subject field around core analytical themes?

But how much of an advance has been achieved for economics if what can be discussed more precisely relates to life situations that actual economic agents would not recognize as their own? The Anglo-Irish marginalists certainly expressed concern with knowingly accepting such a trade-off. Marshall (2013 [1890/1920]: 297), the grand synthesizer of marginalist thought, said that all was well and good with stipulating a problem in abstract mathematical terms, but unless a convincing form of prose could then be found to show how the mathematics described real-life economic encounters there was no justification for believing that the problem under review was genuinely an *economic* problem. Economics, in this sense, needed to remain grounded in the experiences that people could actually be expected to have, and if the narrative content of market models failed such a test then it was difficult to know exactly what they were models of. The progression from the second M to the second S of Samuelson's beloved SMMS narrative therefore appears less a development of what went before than its complete dismissal. Recodifications Samuelson-style can seemingly involve wiping the slate clean. He argued for the unification of economics using mathematical methods to examine the logical properties of maximum conditions. The marginalists' desire for continued grounding in something beyond the hypothetical scenarios of the model world thus no longer had to be a worry. The question of what Samuelson brought to economics should therefore always be asked alongside what he took out of it.

This question is not permitted under an overly linearized history of economic theory, the type in which Samuelson himself engaged. On these accounts anything that was driven out from economics deserved to go as inferior forms of knowledge are made to cede precedence to something more appropriate (Samuelson 1987: 52). Samuelson's tendency to link economists from very different time periods in long hyphenated constructions – such as "Minkowski-Ricardo-Leontief-Metzler matrices" (Samuelson 1962: 1) or "Smith-Allyn Young-Ohlin-Krugman trade paradigms" (Samuelson 2004: 143) – consciously promotes the idea that the evolution of economic theory points inexorably towards its future self. He warned against reconstructing the work of predecessors on the basis of "formalisms that are *not* there" (Samuelson 1998b: 333, emphasis in original), but the suspicion is that he overstepped this mark on countless occasions in the interest of self-certified appeals to scientific unification. As a result, many have pointed to a rather glaring absence of substantive content in Samuelsonian economics, accentuating Robbins's prior promotion of the hypothetical over the empirical (see Chapter 5).

These are the issues around which the following discussion will be based, for they have obvious implications for economics imperialists' attempts to use mathematical market models to explain non-market phenomena. The chapter now proceeds in four stages. In section one, I explore the advantage that Samuelson held over almost all other economists who considered themselves to be methodological trailblazers. There were, in effect, two Samuelsons: the theorist and the textbook writer. The former consistently pushed the boundaries of incorporating mathematical methods into economics; the latter translated these advances for a mass audience as if they amounted to the sum total of economists' concerns. Sections two and three review his attempts to combine these elements to take Robbins's instinct towards social science separatism to its logical conclusion. An economics based on mathematical techniques, Samuelson argued, could be an economics free of all arguments grounded in psychology. Robbins had wanted to emphasize the consequences of conduct rather than its motives, but his attempts to remove utility concerns from discussions of behavioural choice foundered on his failure to adequately rework the concept of rationality as a purely formal mathematical relation. Samuelson went much further. Ultimately he failed to banish utility theory entirely, but not before he had changed the meaning of rationality to render it motiveless as an aid to mathematical tractability. Section four shows how this was necessary if Samuelson's mathematical treatment was to work, but the mathematics itself serves merely to situate all economic activity far away from anything that could be experienced in actual market environments. Despite his assertions to the contrary, this looks like a fundamental ontological break with the world beyond the model, searching for formalisms that are present neither in the work of his predecessors nor in the world they wrote about.

THE TEXTBOOK AND THE THEORETICAL SAMUELSON

Unlike most leading US economists of his generation, Samuelson largely resisted the lure of Washington. Offers were made which would have enabled him to serve in a frontline political capacity, but he avoided risking his academic reputation by tying it too closely to the fortunes of any particular administration (Frenkiel 2015: 84). He even went as far as to announce his indifference to the identity of the country's economic lawmakers, as long as "I can write its economics textbooks" (cited in Nasar 2011: 409). Political influence could be more usefully exerted from a distance, according to Samuelson, by having your arguments made for you by like-minded people who you had taught how to think. Students who had experienced principles based learning through exposure to his simply but rather immodestly titled textbook, *Economics*, would be able to police the boundaries of political discourse so they remained consistent with where economic theory positioned the outer limits of human possibility. By the time of its third edition, Samuelson (1955: viii) felt able to reflect on the rewards of conditioning the thinking of future political gatekeepers: "Contact with hundreds of thousands of minds of a whole generation is an experience like no other that a scholar will ever meet".

The self-styled selling point of *Economics* was that it "concentrates on the big and vital problems" with which policy-makers are required to wrestle if they are to gain social legitimacy. However, Samuelson (1955: v) seems to have protested too much when declaring that such a focus was accompanied by "no single Great Message". There is certainly nothing more than implicit alignment with any of the various policy programmes that came to prominence during the book's lifetime, and he appears to have constructed carefully cultivated ambiguity about his own politics so that a change in administration would not undermine his sales. He described himself rather enigmatically as a "right wing ... New Deal economist" (cited in Leapard 2013: 356), objected to the way in which throughout the 1970s and 1980s "mainstream economics has been moving a bit rightward" (Samuelson 1983b: 790), took pleasure in the fact that those who formed this later vanguard believed the liberal credentials of the 1948 first edition made him a champion of "heresies", and marvelled about how the same musings positioned him amongst progressive critics as "the personification of what was bad about the running jackals of capitalism" (Samuelson 1997: 158, 159). Yet something rather interesting emerges in how Samuelson chose to imprint his own reasoning on the minds of future generations; the content of the learned intuitions he sought to impart changed really rather fundamentally over time (Skousen 1997: 137). However, the Great Message concerned economic theory rather than economic policy.

The early editions of *Economics* focused on macroeconomic imperatives of business cycle management in the service of avoiding future depressions. As times changed, though, and the political system began to reflect new economic priorities, Samuelson's microeconomic voice increasingly came to the fore (Fusfeld 2002: 200). Explicit value judgements continued to be suppressed in the constant avowal that economics was above that kind of thing, but a modern secular theology of social progress through creating robust market institutions made large strides towards the surface of the text (Nelson 2001: 17). Later generations of students were exposed increasingly to the possibility that all economic problems could be stated as market pricing problems: identifying the correct market price signals always revealed the one-and-only truly economic solution (Puttaswamaiah 2002: 8). The contents of *Economics* therefore appear to have fallen increasingly under the spell of the scientific unification that was latent in *Foundations*. Samuelson focused ever more attention on using core economic axioms as the explanation for *all* economic, social and political behaviour, as if a market model was all that anyone ever needed.

In this way, later editions of *Economics* led the trend for textbook framings of the subject field to follow the Robbins (1984 [1935]: 16) definition of an individual acting solely to match available resources to competing ends under conditions of scarcity (Hodgson 2008: 107). Samuelson (1983b: 792) did not rate Robbins's 1932 *Essay* as a methodological treatise, showing barely concealed contempt for its "Kantian a priorism". Yet little separated their respective presentations of what economics should be about, only the route to getting there. The 1961 fifth edition of *Economics* appears to have marked the watershed in this regard. This was where Samuelson (1961: vii) first signalled his desire to write out of the book the discursive tone that helped the reader to bring to mind the image of an often-beleaguered policy-maker trying to balance competing claims on the economic surplus while ensuring that social welfare considerations remained the priority. In its place was substituted a more formalized structure based on underlying propositions designed to allow students to work through the logic of scarcity axioms (Thompson 1999: 224). The image of the policy-maker beset by real-life struggles accordingly faded very far into the background, replaced by the image of a technician capable of unlocking the secrets of the internal workings of markets through applying the most up-to-date microeconomic techniques. By 1961, then, students coming newly to the subject field via *Economics* were taught instinctively how to think in relation to the world within the market model and not in relation to the world beyond.

Why, though, did it take Samuelson so long to settle on the presentational formula that defines the later editions of *Economics*? This was, after all, only to bring his way of addressing students into line with how he addressed his professional peers. The authority he had already forged to speak on behalf of his subject is what brought him to the attention of textbook publishers in the first place. He was never shy in making his case, being blessed with "world-girdling ambition" (Warsh 1993: 184). His path-breaking accounts of how to do economics stood out most obviously for their initial forays into what would later become a fully-fledged formalist turn. The early Samuelson (1938: 64, 70) pressed ahead in a much more thoroughgoing manner than had ever previously been achieved in presenting mathematical relationships as direct synonyms for economic relationships (see also Samuelson 1986: 797). The discursive tone of the first four editions of *Economics* therefore looks like something of a detour from the agenda previously set out in *Foundations*. It was offered to his professional peers as an indication of what needed to be cleared out of the way before mathematical economics could take centre stage (Samuelson 1983a: xxv, 1986: 802). But it required at least four rewrites before *Economics* caught up in matters of form.

Samuelson conceived of mathematics as a logic that enabled useful generalizations of what economic conduct would look like in timeless conditions in which historical specificities could be assumed away (Kirman 2011: 128). Some such idea was certainly in vogue when he began his doctoral studies in the late 1930s due to the ongoing reappraisal of the work of Vilfredo Pareto (Samuelson 1998a: 1381). Following his embrace of ordinalism, Pareto (1971 [1906]: 120) informed his reader that "the individual can disappear" from the theoretical activity of determining an equilibrium, "provided he leaves us a photograph of his tastes". This seems to have been a decisive turning point for Samuelson, as it allowed economists free rein to remove the interfering influence of psychology so that all explanatory content was concentrated in the mathematical structure of maximization problems (Amariglio & Ruccio 2001: 144). However, those who are more attentive to Pareto's arguments post-*Manuale di Economia Politica* (see Chapter 5) have asked whether his apparent dismissal of the individual is less a theoretical recommendation or a warning of the ontological costs involved in allowing economic theory to be subsumed by mathematical formalism (Soto 2009: 53). Tracing the behavioural path implied by a mathematical equation is one thing, but it is altogether another to do so at the expense of interest in the real protagonists of social processes (Raico 2012: 8).

This produces a notable tension that Samuelson ignored in the first four editions of *Economics* and assumed was irrelevant from edition five onwards. How might economic theory proceed if its newest recruits are to receive instruction, as per Samuelson's earliest wishes for *Economics*, on the dilemmas that the real world throws up for policy-makers (see Samuelson 1955: vi)? The answer seems initially to have been that theory and instruction were to occupy parallel universes, positioned in conflicting ontological realms. Jevons's (1876: 624) attempted resolution of the Anglo-Irish *Methodenstreit* therefore appears to survive in the early work of the two Samuelsons (see Chapter 4). Jevons's compromise was that there were simply two different tasks, one of formal abstract theory and the other of contextual historically informed theory, both of which were crucial to the overall development of economics. Historically grounded and culturally socialized economic agents were cast out of Samuelson's *Foundations* as mathematical techniques were used instead to explore the underlying logic of purely abstract economic behaviour, and yet still they featured prominently in how students learned the subject field through the first four editions of *Economics*. The recombination of the dominant voice across the two texts only occurred with Samuelson's eventual construction of a more formally oriented *Economics*.

From this point on, the prior tension became much less noticeable because by now it was argued that obedience to abstract mathematical theorems was all that was necessary for either theoretical or instructional purposes (Davis 2011: 49). Pareto had suggested an alternative destination in sociology for historically grounded and culturally socialized economic agents if no role could be found for them in the purely logical domain of economic theory (Barruchello 2012: 166). His mechanical conception of agency under purely logical decision-making had an important counterpart in the synthetic agent that inhabited his sociological theory of non-logical decision-making. Pareto's *Homo economicus* thus had its social consciousness removed and placed elsewhere (McClure 2010: 643). Samuelson's *Homo economicus*, though, was provided with no escape route for its socially aware alter ego.

Samuelson therefore appears to have consciously placed economic theory in the grip of a Paretian dystopia. Vincent Tarascio has undertaken a careful study of Pareto's evolving concerns about the direction in which mathematical formalism would take economic theory, a nuance that was lost on both Robbins and Samuelson as they prepared the field in their different ways for its eventual takeover by formalist techniques. Tarascio (1972: 415) argues that one of Pareto's major methodological achievements was to remove "the façade to expose so-called 'economic man' for what he was – neither economic nor man". Nicholas Georgescu-Roegen (1966: 104) suggests that Samuelson failed to take heed of Pareto's counsel. "[F]or a science of man to exclude altogether man from the picture is a patent incongruity", he argued. "Nevertheless, standard economics [in the Samuelsonian mould] takes special pride in operating with a man-less picture … [It] requires a computer not an agent." Pareto thus appears simultaneously in irreconcilable guises in the prehistory of economics imperialism. The Pareto of the Pareto specialists seems to have done most to penetrate the sugar-coated account of the marginalist revolution and to raise significant concerns about what mathematical formalism would subsequently do to economic theory (Drakopoulos 2012: 548). There is no obvious justification for economics imperialism here. However, Robbins and Samuelson both seem to have found something different in Pareto. They saw his work as the precursor to cleansing economic theory of everything except a strict mathematical logic (Hodgson 2001: 233). Samuelson's approach to the history of economic thought is once again important. What matters, he argued, is what today's economists find useful in their predecessors' work, not what those predecessors thought they were doing for themselves (Samuelson 2009: xii).

Samuelson thus seized upon and worked with the one thing that appeared to make Pareto nervous about his own economic theory: namely, how the embrace of the purely logical domain equated to an act of confinement (McClure 2005: 611). Within that domain, Pareto's mathematical instincts had shown that systems of equations could capture in abstract terms the reciprocal relations that interacted to produce points of economic equilibrium (Tarascio 1974: 362). Yet there was something specific about Pareto's approach. It conceived of equilibrium via the relationship between things rather than the relationship between people. Once the significance of the economic agent was reduced to the clues that they had disclosed about the structure of their tastes, it is the things on which those tastes were imprinted that mattered most. What then comes to the fore in subsequent economic theory is how those things relate to one another as other inanimate objects do. Pareto pointed out, as a consequence, that his equations of equilibrium were derived directly from the equations of rational mechanics: "they are old friends", as he put it rather endearingly (Pareto 1999 [1900]: 250). Samuelson felt no embarrassment in pushing the mechanical analogy until economic theory had been wholly submitted to it (Mirowski 1991: 223).

SAMUELSONIAN ECONOMICS AND THE ECLIPSE
OF PSYCHOLOGY

Samuelson found no need to bother himself with Pareto's contortions about whether indifference curves were still experimentally derivable in principle even if there were no suitable experiments to show in practice how this was so (Lenfant 2012: 148). He plumped straightaway for what he thought was the more easily mathematicized of Pareto's two reworkings of Francis Ysidro Edgeworth's (2003 [1881]: 28) original account of indifference curves: namely, that they represent market choices in the presence of stabilized economic behaviour (Giocoli 2003: 72). The target of stability equations, though, differed between Pareto's and Samuelson's approaches. For Pareto it meant observations of repeated patterns of choice to be rendered meaningful as statistical objects, whereas for Samuelson it meant potentially repeatable patterns of choice as captured solely by the mathematical logic of Lagrange multipliers. The former was after the fact of decisions having already been undertaken (consistent with Hilbert's concrete axiomatization), whereas the latter was before that fact (consistent with his formal axiomatization). With Samuelson's total subordination of economic theory to mechanical analogy, it was unnecessary to wait until the choice had been made to know what it was going to be. Mechanical systems, after all, are established on the inherent predictability of a single efficient path to an optimal solution (Jarvis & Mosini 2007: 61).

At least for a short period, the early Pareto insisted that there was no need to measure anything directly to preserve the integrity of core economic propositions (Chipman 2010: 18). "[I]n reality", he wrote (Pareto 1999 [1900]: 248), "nobody is capable of measuring pleasure". Pareto had hinted at the possibility of rebasing economics on ordinalist scales of preference right from his very earliest work (Weber 2001: 544), substituting his index of pleasure for Jevons's hedonistic calculus and removing all need to establish utility functions on the real number line (Bruni & Guala 2001: 36). Pareto was helped in this regard by being able to call upon advances in number theory that were unavailable to the first generation of marginalists. Ernst Schröder's (1873) pioneering elaboration of the distinction between cardinal and ordinal numbers was published two years after Jevons's *Theory of Political Economy* and Menger's *Grundsätze der Volkswirtschaftslehre*. Georg Cantor's (1883) formalization of that distinction was another ten years in the making, but it was only then that ordinal numbers truly began to enter the consciousness of mathematicians, let alone economists. At that time economists were busy watching Menger and Schmoller fight the increasingly vicious Austro-German *Methodenstreit*.

The use of ordinal number theory allowed for the adoption of mathematical forms through which economics could dispense with specific utility functions (Lewin 1996: 1302). This did not lead, however, to a once-and-for-all excision, not even in Pareto's own work. The textual evidence shows that he was less confident in an economics founded on ordinalism in 1908 than he was in 1901 (Weber 2001: 561). The more in-depth his sociological studies, the more it looked that his conversion from direct measurability of utility was a temporary three-year interlude from 1898 (Wood & McClure 1999: 8). Samuelson, though, was not to be dissuaded. He was adamant that better mathematics of the sort first hinted at by the early Pareto provided the ideal medium through which the common core of all economic problems could be expressed. This was more than merely the repetition of the early marginalists' mathematical intuitions, involving instead actively demonstrating how it might be possible to realize the unification of all relevant questions within economic theory under the maximization principle. A surgical strike on utility theory, he believed, would let increasingly sophisticated mathematical techniques do what they do best in allowing for more precise specifications of the world within market models.

Samuelson's proposed solution passed through the post-Paretian indifference curve approach to consumer behaviour pioneered in the early 1930s by the English economists John Hicks (1934, 1939) and Roy Allen (1934a, 1934b). Both were employed at the London School of Economics, lending extra weight to Robbins's (1932: xlii, 1971: 11) assertion that his scarcity definition was not his alone, so much as something more generally in the air around him (see Chapter 5). When Robbins said that he was merely capturing the advances already being made by the most talented economists of his generation, he would certainly have counted Hicks and Allen prominently amongst their number. Samuelson (1972: 255) noted in his Nobel Lecture that in the intellectual environment pre-Hicks and Allen, "utility theory showed signs of deteriorating into a sterile tautology. Psychic utility or satisfaction could scarcely be defined, let alone be measured." Utility functions are *post hoc* reconstructions of how the individual might be understood to have behaved were the surrounding theory to be true, but the only reason to impose such motives on consumer behaviour in the first place is the desire to believe that this is the case (Blatt 1983: 170). Once more we are drawn to the possibility that the relevant representational relationship is between the market model and economic theory, not between the market model and the world it supposedly resembles (see Chapter 3). There is clear circularity here, and Hicks (1934: 61) and Allen (1934b: 203) tried to cut through it by resetting the underlying presumption on which the theory was based. They moved away from assuming that utility was cardinally measurable to assume instead that it was something that one might simply recognize having more or less of. There was no longer a need to know how much of it was being enjoyed at any particular moment of time or even the psychological form such enjoyment took. This reduced all analyses of consumer behaviour to individuals who had no historical or cultural reference points other than themselves at prior points in their life (Pollak & Wales 1992: 91). What was now important was how individuals used their available resources to reach the highest attainable indifference curve relative to their current preferences (Blaug 1997: 328).

Samuelson (1938: 61) was impressed with the general "discrediting of utility as a psychological concept", but he did not think that Hicks and Allen had gone far enough. Utility theory, albeit in ordinal rather than cardinal form, was still present within the basic structure of propositions that guided their indifference curve analysis, as was the accompanying circularity in positing the market model that is presupposed by utility theory as the best demonstration of how utility might be maximized. Samuelson wished to eliminate both altogether. His first published work on the subject appeared in print exactly 200 years after Daniel Bernoulli (1954 [1738]) had initially used the concept of utility to describe what lay behind individual preferences. Samuelson appropriated what in other circumstances might have turned into bicentennial commemorations of an economic master concept to declare that enough was finally enough. He was concerned that, even in its reformulated state post-Hicks and Allen, utility theory was unable to yield sufficient empirical content to enable economics to become a genuine observational science (Samuelson 1938: 63). Samuelson had noticed that Gustav Cassel (1967 [1918]) had been successful 20 years earlier in simplifying Walras's equations of general equilibrium by extracting the concept of utility from them (Walker 2011: 291). Knowing that Hicks in particular had been responsible for bringing Walras's economics to the attention of the English-speaking world (see Hicks 1984: 285), Samuelson set out to replicate Cassel's achievement but this time on Hicks's (1934, 1939) and Allen's (1934a, 1934b) work. In that way, he argued, it would be possible to remove from economics the last "vestigial traces of the utility concept" (Samuelson 1938: 61).

Samuelson's unapologetically whiggish approach to the history of thought led him to believe that behavioural theorems evolved through shedding assumptions that had been crucial to earlier intuitive theoretical propositions but had latterly been shown to be formally unnecessary (Rizvi 2007: 378). He saw his task as merely the next stage of the collective process through which economic analysis might be unified around the smallest number of essential axioms, with utility theory decidedly not being one of them. "From its very beginning the theory of consumer's choice has marched steadily towards greater generality, sloughing off at successive stages unnecessarily restrictive conditions. From the time of Gossen [the 1850s] to our own day [the 1930s] we have seen the removal of (*a*) the assumption of linearity of marginal utility; (*b*) the assumption of independence of utilities; (*c*) the assumption of the measurability of utility in a cardinal sense; and (*d*) even the assumption of an integrable field of preference elements" (Samuelson 1938: 61). Samuelson's criticism of Hicks and Allen was that they had not seized the mantle of this historical trend to make the final break with utility theory as a whole. He believed that he had the mathematical tools to accomplish such a goal, thereby approaching the issue of behavioural dispositions "directly", instead of taking the indirect route through Hicks's and Allen's indifference curve analysis and its starting point in ordinal utility theory (Samuelson 1938: 62, 71).

Samuelson thought that a more direct approach was possible simply by observing what people had already done. This was the basis of his famous revealed preference theorem, which often continues to be treated as foundational for subsequent advances in precision of mathematical market models. It is permitted to occupy such rarefied status because it is widely assumed to enable behaviour to be captured as a purely theoretical matter (Rugina 2005: 234). Samuelson's "directness" when it came to behavioural issues is therefore merely a theoretical route to what his mathematical models would allow him to say, and his "observations" likewise relate to what his mathematical models would allow him to see. The words that he used often give the impression that a moment of ontological unification is in sight because they seem to refer to the world beyond the model, but first glances can deceive. In these instances

– in many others too – he appealed to economic vocabulary we are most familiar with in relation to real phenomena, but to describe the mathematical features of the world within the market model. Consequently, he narrowed his horizon at most to moments of derivational unification (see Chapter 3).

The most important aspect of the revealed preference theorem for current purposes is that it removes all social stimuli from the discussion of how economic agents choose to act (Rosenberg 1992: 138). We no longer need to know anything about people's thought processes to account for their conduct: not what sort of person they are, not how they are trying to imprint themselves on the world, not which worlds would meet with their consent. Samuelson's revealed preference theorem appropriates Robbins's desire to remove the need to ask "why" questions when undertaking research into economic behaviour, and in its place he introduced a purely logical conception of rationality. For anyone to be presented as a suitable subject for mathematical economics – what Samuelson (1938: 63, 1948: 243) calls "our idealised individual" or "the individual guinea pig" – they merely have to be consistent in the choices they make. Nothing else needs to be known about them except that their choices would obey the purely logical conditions of reflexivity (written formally, $\nabla x \in X: xRx$), completeness ($\nabla x, y \in X: xRy \lor yRx$) and transitivity ($\nabla x, y, z \in X: (xRy \,\&\, yRz) \rightarrow xRz$). These are the situation-defining assumptions that identical combinations of goods are assumed to be of equal quality, different bundles are directly comparable, and consistent ordering prevails across any combination of two bundles.

Samuelson used his logical definition of rationality to demonstrate what could be known having first stipulated that economic behaviour is oriented around equilibrium conditions. The ratio of prices drawn through the equilibrium point creates a simple feasibility set that can be defined mathematically and visualized diagrammatically, within which every combination of goods could have been chosen instead of the equilibrium bundle. "But they weren't", he added, simply. "Hence, they are all 'revealed' to be inferior" to the combination represented by the equilibrium point. "No other line of reasoning is needed" (Samuelson 1948: 244). Samuelson thus proposed to change the focus from theorizing agential motivation to working through the implications of hypothetical conduct under conditions of constrained choice (Leapard 2013: 356). The hypothesis that a logically rational person will always choose one bundle of goods over another when they have already shown themselves to prefer that bundle certainly qualifies as a testable hypothesis (Backhouse 1994: 214). However, Samuelson seems to have treated the mathematical procedure through which the revealed preference theorem operated as both hypothesis *and* its accompanying test. As a route to his desired destination of an increasingly rigorous mathematical economics, the revealed preference theorem circumvented questions that might be answered empirically (Mulberg 1995: 65).

Samuelson's (1983a: xviii) objective was to rid economics once-and-for-all of the "calculus corsets" bequeathed by the marginalists. Not only did he think through the problems of economics mathematically, but also the mathematics came first. Better mathematics produces better economics was his motto, almost to the point at which the mathematics got the economics it deserved. The problem with differential calculus was not epistemological, because Samuelson remained wedded to the application of his simple Lagrange multipliers even as other, more mathematically gifted economists took the subject field in more complex analytical directions (see Chapter 7). How could this be the problem, as the very notion of an equilibrium point that defines as inferior all alternatives to the revealed preference is grounded precisely in the intuitions of a differential calculus mindset? Exit the mindset and the revealed preference theory becomes inoperable. Samuelson never lost faith in the mathematical functions that can be explored through many orders of partial derivatives, and he never sought anything more complicated for his economic theory. His objection to economists' prior uses of differential calculus was instead ontological, because he assumed they readmitted the psychological principles of early Anglo-Irish marginalism. Samuelson (1948: 244) instead found salvation in the use of contemporary index number theory (Ramrattan & Szenberg 2019: 141–2), a mathematical technique that demanded no more for its translation into economic theory than that behaviour was consistent (Samuelson 1998a: 1380).

THE MISDIRECTION IN SAMUELSONIAN EMPIRICS

In later life, Samuelson (1998a: 1381) wrote that his brute behaviouralist programme of revealed preference achieved "all I hoped for". Others, however, have announced its almost complete failure (Rutherford 2011: 316). Samuelson based his work methodologically on the need for what he called operationally meaningful theorems, which he defined in *Foundations* as "simply a hypothesis about empirical data which could conceivably be refuted, if only under ideal conditions" (Samuelson 1983 [1947]: 4). This looks to be an early pre-emption of Robert Sugden's (2000: 24, 2009a: 17) credible-worlds thesis, in which the very possibility of being testable is an acceptable alternative to actual testing (see Chapter 3). The reference to "if only under ideal conditions" is pointing towards a thinkable, even if wholly unlikely, state of affairs. Why Samuelson believed his theory provided something that was thinkable but Jevons's and Robbins's did not is something of a mystery, because each in their own way abided by the criteria of credible worlds. Yet in a swipe at Jevons and Robbins respectively, he wrote that his operationally meaningful theorems "are not deduced from thin air or from *a priori* propositions of universal truth and vacuous applicability" (Samuelson 1983 [1947]: 5). His critics are not so sure.

The consensus amongst intellectual historians today is that Samuelson smuggled back into his revealed preference theorem both the instincts and the image of utility theory (Caldwell 1982: 197). His intention was to investigate "*refutable* hypotheses [related to] the observable facts on price and quantities demanded" (Samuelson 1972: 256, emphases in original), but ultimately his demand curves took their shape for the same reasons as Hicks's and Allen's. He confessed that his own "solution curves are the conventional 'indifference curves' of modern economic theory" (Samuelson 1948: 245), and as Hicks and Allen had derived those on measures of ordinal utility then, by implication, Samuelson must have done so too. His charge of circularity against all his predecessors could therefore equally have been levelled against his own mathematical recodification.

Stanley Wong (2006 [1978]: 88) has argued that there is only one implication of Samuelson's (1950: 370–72) eventual admission of the observational equivalence between his own and earlier theories of consumer behaviour: that there is no real "problem to which the revealed preference theory is a proposed solution". Samuelson had begun in 1938 by claiming that he had secured a new basis for economic theory requiring no reference to utility in any form; by the mid-1940s that the revealed preference theorem allowed an indifference map to be constructed entirely separately from any discussion of economic motivation; but by 1950 only that the revealed preference theorem mimics in observational terms ordinal utility theory (Ruccio & Amariglio 2003: 99).

If revealed preference and ordinal utility are not just observationally but also axiomatically identical, there is nothing the former can enable the theorist to do that the latter cannot (Sen 1973: 243).

Samuelson dug his heels in, though. He defended his achievements by suggesting that they belonged to something rather larger than economic theory, insofar as they represented the "marriage between Haberler–Konüs index number theory and Gibbs finite-difference formulations of classical phenomenological thermodynamics of the 1870s" (Samuelson 1998a: 1380). Haberler–Konüs is another of Samuelson's hyphenated constructions, invoking Gottfried Haberler's (1927) work on index numbers and Aleksandr Aleksandrovič Konüs's (1939) early application that allowed the cost of living to be compared over time. Meanwhile, the Gibbs (2014 [1902]) measure of the probability of a system being in a particular state was crucial for the derivation of all sorts of difference equations. Samuelson believed that index numbers contained useable information about maximum states but remained resolutely ordinal in nature, meaning that nothing had to be directly observed, measured or calculated before being added to Gibbs-type equations to demonstrate the theoretical breakthrough contained in his revealed preference theorem.

However, Hendrik Houthakker (1950: 161) showed that the mathematics did not work in the way Samuelson claimed. The path to his desired solution required the addition of situation-defining assumptions familiar from Hicks and Allen, thus moving him back onto intellectual territory he said he had left behind. Samuelson readily acknowledged the role that Houthakker had played in allowing him to address what he presented as a minor early oversight, but he was adamant that once corrected this allowed him to argue even more forcefully that he had been right all along about the radical break established by his revealed preference theorem (Samuelson & Swamy 1974: 580). Willard Gibbs, he pronounced, "led me to the promised land before there was a promised land" (Samuelson 1998a: 1381). For his part, Houthakker saw matters very differently. Samuelson was insistent that nothing had changed in eventually admitting that the more sophisticated mathematics of the revealed preference theorem occupies exactly the same intellectual terrain as the utility theorists of his day. Houthakker was just as insistent that nothing could ever be the same again for Samuelson's theory after such a climbdown: "The stone the builder had rejected in 1938 seemed to have become a cornerstone in 1950" (Houthakker 1983: 63).

It is interesting that Samuelson chose to defend his economic theory mathematically rather than economically. To say that his defining contribution was to insert a mathematical technique for deriving index numbers into another mathematical technique for adjudicating on the likelihood of encountering a maximum condition is only really to highlight the absence of any obvious substantive economic content. This is clearly uneconomic economics (Watson 2018: 83). Samuelson's stated aim was to make mathematical economics as robust in observational and operational terms as mathematical physics (Sent 1998: 44). However, when David Hilbert (1924) moved from pure mathematics to mathematical physics it was by heading in the opposite direction to Samuelson, away from existence proof-making and towards more concrete forms of axiomatization. To use something similar to Samuelson's hyphenated name chains against him, Fourier–Frege–Lesbegue–Cauchy scepticism about existence proofs that were unmoored from anything real eventually left an impression on Hilbert's thinking (see Chapters 1 and 2). The mathematical imagination had to encompass both formal and concrete axiomatization, according to the later Hilbert, and the trick was to be sufficiently discerning to know which subject matter required which type of axiomatization (Hilbert & Bernays 2003 [1934]: 2). Hilbert (1967 [1925]: 376) thought that mathematical physics could reach its full potential only in the context of "*inhaltliche*" (or contentual) statements. This required content to be assigned to the model's primitive features from directly describable experiences before any hypothetical mathematical modelling could proceed. Samuelson wrote of his ambition to emulate the successes of mathematical physics, but he departed significantly from Hilbert's methodological prescriptions for how best to do so.

Mathematical physicists specify calculable results from their theories, enhancing their professional credibility by showing how these results survive empirical testing through observational data. That is how they win their Nobel prizes (Dardo 2004: 33). Yet there is nothing like this in the Samuelsonian tradition. Given his epistemological strictures about operationally meaningful theorems, it is remarkable that the systems of equations which provide *Foundations* with its explanatory content stand entirely by themselves and have to be accepted on their own terms (Samuelson 1983 [1947]: 7, 1963: 233, 1983b: 791). Samuelson's mathematical economics is a purely logical pursuit (Gillies 2004: 189). There is no instance where a calculable result is derived within a market model that might then be taken out into the world to see how well it performs empirically. Little wonder, then, that Wong (2006 [1978]: 51) doubted whether there was a genuine *economic* puzzle to which Samuelsonian economics responds. Creating a hypothetical mathematical model to explain how its internal features shed light on abstract economic theory is very different to creating a hypothetical mathematical model with in-built inferential capacity to investigate how real phenomena are related.

Samuelson (1983 [1947]: 9) has to allow the simultaneous determination of all variables at the moment he imposes the conditions of equilibrium, otherwise his index number theory does not work to demonstrate the mathematical features of maximization. As George Brockway (2001: 346) suggests, all the variables in his theoretical system consequently "become like the notorious arrow in the paradox of Zeno of Elea, which cannot move because it is where it is and not somewhere else". The maximum conditions that act as solutions to Lagrangian functions must remain purely abstract entities which necessarily escape the pressures of falsification. There is always insufficient data to know for sure under marginalist mathematics whether anything is being maximized, but Samuelson accepts an even more fundamental level of uncertainty. In his approach, nothing ever gets tested in a conventional empirical sense, and therefore even the *desire* to know for sure whether anything is being maximized in practice is absent (Hodgson 2001: 239). Yet as Deirdre McCloskey (2005: 76) notes, to say that an effect has a potential mathematical character suggestive of maximization is not the same as having detailed knowledge about that effect. Despite their name, existence proofs do not tell us what actually exists in the world but limit us to talking about what exists in the relationship between the assumptions on which the model world is based (see Chapter 3). The Samuelsonian recodification of economic theory therefore seems to be merely a method for deriving conclusions from assumptions.

Samuelson (1952: 58, 1972: 12, 1977: 47, 1983 [1947]: 12, 1983a: vi) insisted throughout his career that mathematics was a language like any other: what was said mathematically was only what could be said using prose but in a more compact, robust and aesthetically pleasing manner (see Chapter 1). However, it is inconceivable that anyone would set out *wanting* to prove the existence theorems that have come to dominate economics post-Samuelson had they been restricted methodologically simply to the use of prose. Why would the desire take hold to talk about the world in ways that were known to be unrealistic if the justification for doing so had to rely on words alone? Samuelsonian mathematics has therefore changed much more in economics than merely how economists might say what they would always have wanted to say in any case. It has also shifted the very act of what might be considered thinkable, so that increasingly the words that explain the relationships contained within the mathematical symbols reveal nothing beyond internal aspects of the accompanying systems of equations. It is possible to get away with significantly more self-referential theorem-building when dressing it up in the ostensible rigour of abstract mathematics.

"For few commodities", wrote Samuelson (1983 [1947]: 258), "have we detailed quantitative empirical information concerning the exact forms of the supply and demand curves even in the neighborhood of the equilibrium point". He continued that "this is a typical problem confronting the economist", before then offering a purely mathematical solution to what is clearly an economic problem. "[I]n the absence of precise quantitative data [the economist] must infer analytically the qualitative direction of movement of a complex system." In other words, the lack of good empirical evidence relating to the internal dynamics of market models does not compel the need to go out and generate such data. Deriving mathematical structures consistent with qualitative theorems does the job in a roundabout manner by ensuring that the question of empirical content does not need to be asked in the first place. "We have merely to show", Samuelson (1983 [1947]: 39, emphases added) asserted, "that a wide variety of economic problems can be so formulated as to yield a conclusive determination of the *sign* of the criterion". Here, he is saying that the use of index number theory successfully exorcises all aspirations towards using numbers in their cardinal sense. Something significant happens, though, to a supposedly observational science in which no actual observation ever needs to take place. The rhetoric of Samuelsonian mathematics changes from "how much" to "whether". The narrative content of market models overreaches if it says anything more than whether an equilibrium exists as a maximum condition and whether the mathematical equations specify the dynamic path to a steady state (McCloskey 2005: 76). Samuelson thus reduced economics to the abstract study of direction of change in decision variables following a shift in the parameters of the economic system (Silberberg 2005: vii). There are good reasons to doubt, though, what can truly be said about real-world causality from this perspective (Dopfer 1989: 50).

If this were not enough, there are important admissions in both *Foundations* (Samuelson 1983 [1947]: 22) and his Nobel Lecture (Samuelson 1972: 258) that not all economic problems fit the maximization frame. It is not, he wrote in *Foundations*, "the 'open sesame' to the successful unambiguous determination of all possible questions which we may ask" (Samuelson 1983 [1947]: 41). However, in laying the basis for what he thought economics should be in the future, he argued "that *everything* interesting is contained in the inequalities associated with an extremum position" (Samuelson 1983 [1947]: 231, emphases in original). Maximization devices, he began his Nobel Lecture, provide "a better, more economical, description of economic behaviour", even if what was being observed was not actually a maximum condition (Samuelson 1972: 251). Samuelson's commitment to the idea that all problems might usefully be reduced to the analytical properties associated with systems exhibiting maximizing tendencies ensured that the character of real phenomena was never the basis for a valid counterargument.

He was explicit on this point: "some problems which do not appear to involve extremum positions can at times be converted into an equivalent maximum or minimum problem … [T]here is not much which cannot be brought under this heading" (Samuelson 1983 [1947]: 52, 21). He continued that "the conversion of a problem whose economic context does not suggest any human, purposive, maximizing behavior into a maximum problem is to be regarded as merely a technical device for the purpose of quickly developing the properties of that equilibrium position" (Samuelson 1983 [1947]: 53). This looks very much like the manifesto for a strategy of derivational unification, but of course the philosopher Uskali Mäki (2001a: 493–8, 2009a: 363–4) asks us to think about it as *mere* derivational unification to distinguish it from something more significant in explanatory terms. It is clear that the mathematical structure is being given priority here and that representations of social reality are made to conform to that structure. If we scratch the surface even very gently here, a pre-emptive manifesto for economics imperialism also becomes apparent, even as it suggests that the imposition of the market model onto non-market contexts is limited in explanatory terms to the mathematical contrivances of derivational unification. "Thus", wrote Samuelson (1983 [1947]: 23) in what might be the most instructive sentence in the whole of *Foundations*, "we really argue backwards from maximizing economic behavior to the underlying physical data consistent with it".

TAUTOLOGICAL APPLICATIONS OF MARKET MODELS

Despite Samuelson's (1983a: xxvi) best efforts to create "a general theory of economic theories", his critics are surely correct that there must be more to explaining the outcomes of actually observed economic relations than can be said using mathematics alone. Samuelson's status as the field's most important unificationist is evident in his rock-solid commitment to the idea that a single general argument pattern can explain everything there is to explain. We must be clear, though, that this is only everything within economists' self-made model worlds. The explanatory potential stretches no further than that, because what remains is a very peculiar conception of economic agents. They are willing to conduct themselves at all times in an optimally economizing manner that produces a maximum solution, even if this forces them to act against their considered judgement of what they would most like to do. Their actions bring clarity to the world within the model, but they would struggle to create a meaningful life for themselves in the world beyond. It is certainly possible to conceive of behavioural attributes being limited to automatic rule-following as an abstract thought experiment, but what use is it for solving anything other than artificial problems?

Samuelson himself appears to have become increasingly ambivalent about methodological issues as he aged, saying: "I think most economists, most of the time, do not have to be explicitly aware of methodological questions" (cited in Pizano 2009: 110). But perhaps this is a position he had to take if his peers were to be comfortable with the homogeneity he had imposed upon behaviour. The mathematics simplifies – that is always its job – but here it oversimplifies to such an extent that the people being described by the theory are unrecognizable as people exhibiting a genuinely human form. The Samuelsonian override, as it might usefully be called, strips the economic agent of the degree of consciousness necessary to be anything other than a mindless devotee of a single behavioural rule. The impulse towards economics imperialism today proceeds with at least one foot in these tracks. On so many occasions, non-market contexts are made to confirm to the world within market models by denying individuals the ability to be goal-directed in a manner which puts them in charge of setting the goals, unless we are to assume that their only objective in life is to reflect the mathematical structure's maximization fetish. After all, as Geoffrey Hodgson (1999: 37) notes wryly, "it is easier to demonstrate mathematical prowess on the basis of assumptions of your own choosing".

The Samuelsonian override ensures that to act rationally is to pursue the relevant maximum condition without giving any thought to whether the prize was worth the chase. In trying to avoid complicating factors associated with the social determination of behaviour, Samuelson preserved the pristine nature of his mathematical structure by embracing a conception of rationality that emptied it of all self-perception of purpose. The all-encompassing features of the maximization principle invoke an invariant trajectory law to which all behaviour must necessarily be subsumed (Dopfer 1989: 55). Many economics imperialists today flirt openly with the same excision, putting them in the seemingly paradoxical position of trying to explain social behaviour without reference to the social determinants of that behaviour. The narrative imposed upon the mathematical structure of their market models might attempt to hide this fact, but it is the mathematical content and not the narrative content that ensures the models' tractability. Echoing Robbins's (1984 [1935]: 23–31) initial focus on means not ends (see Chapter 5), being rational is enough in itself. There is no sense of needing to be rational relative to any particular objectives the individual might have decided upon for themselves. Economic agents subjected to the Samuelsonian override do not act at all in any literal sense of the word; they merely act out the theorist's assumptions about the abstract characteristics of the context in which they find themselves (Hay 2002: 104). Agential self-awareness is sacrificed to guarantee that the model always produces a maximum solution.

Samuelson's rationality postulate is therefore what Alan Musgrave refers to as a domain assumption and Uskali Mäki (2000: 324) an applicability assumption (see Chapter 3). This is something that describes the type of behaviour allowed by the model world and not any actual behaviour itself. It is "an *ad hoc* assumption, in Popper's sense" (Musgrave 1981: 381), and it points only in the direction of an increasingly untestable theory. Allan Gibbard and Hal Varian (1978: 665) consequently place Samuelsonian models in the camp of "caricatures". Here, the aim is to use theory to present a storyboard that purports to inform the reader what the world is like, but it does so only by suppressing other, more realistic narrative possibilities. The prescribed narrative is essential to market models' usefulness as reasoning tools, but it is not conditional upon the mathematical structure of the models. It stands alone as a normative account of the theorist's preferred state of affairs (Strassman & Polanyi 1995: 96). Pride of place in this regard for Samuelson goes to pristine market institutions and their ability to deliver on the promise of perfect competition. The mathematical structure of *Foundations* is therefore not just another language for capturing economic generalities (Samuelson 1952: 59), because it enforces market-conforming narratives as the only possible ordinary language form for discussing the challenges of everyday economic life.

However, Samuelson's mathematical structure does nothing to solve the most pressing problem of how economic theory conceptualizes the market as the equilibrium interaction of demand-and-supply dynamics. Without imposing perfect competition as the institutional setting for market-based behaviour, it becomes impossible to derive a supply curve even theoretically, let alone empirically (Hill & Myatt 2010: 112). One half of the basic demand-and-supply diagram therefore does not exist in anything other than the most extreme of special cases (Thompson 1999: 231), even if its absence is masked by the narrative content of most market models. Samuelson, though, needed active resemblance of both halves to be present in his model world if it was to produce operationally meaningful theorems concerning prices, because in seeking to eliminate all discussions of utility from economics he shifted its core from value-theoretical to price-theoretical accounts (Gramm 1988: 226). By way of justification, he told his peers a whiggish intellectual history of truth conquering error and of the consequent development of a higher plane of learning. But a price-theoretical approach to equilibrium conditions presumably leads nowhere if the assumption of perfect competition is maintained in the absence of a derivable supply curve. The symmetrical relationships of perfect competition are raised to a significance that is simply not replicated in the real world, as if each person has internalized the theorist's system of equations and will never consider acting in any other way. The impetus here is merely mathematical expedience, so that Samuelson's chosen systems of equations produce obvious solutions. However, in the bigger scheme of things, accepting clearly unrealistic assumptions about perfect competition simply because they make the mathematics more tractable seems to be an early admission of defeat.

To make any attempt to explain how economies react when out of equilibrium, those working within the Samuelsonian tradition have chosen merely to add a mathematically appropriate difference equation to capture the image of price adjustment relative to the initial disturbance of equilibrium (Boland 2003: 216). As long as the right difference equation is inserted, no shock to the system need ever be big enough to prevent the restoration of equilibrium under the assumption of perfect competition (Keen 2013: 229). This might well tell us that markets are always capable of clearing when initial conditions are suitably specified in line with economists' model worlds, but it says nothing about the much more fundamental issue of how, *in practice*, actual markets work. Even adding an artificially contrived supply curve to augment what can be gleaned from the revealed preference theorem about market demand takes the analysis no further towards what would be most useful to know.

As a result, John Quiggin (2010: 5) has spoken of a "zombie" economics, in which ideas that should repeatedly have been declared dead continue to stalk the land of the living. As discussed at length by his critics, everything which Samuelsonian theory has to treat as an historical abnormality belongs to the world of relatable human experience, but that which it takes to be essential economic behaviour belongs somewhere else entirely. Samuelson's self-made model worlds feature agential characteristics that are only recognisable there. At most they exist at the outer limits of the thinkable as per Sugden's definition of a credible-world model, but they fall well short of anything that might be considered a plausible resemblance to the real world (Mäki 2009b: 40). They become thinkable only when market adjustment is reduced to a Markov process in which future economic behaviour is a function solely of the initial conditions before the market adjustment began (Kirman & Zimmermann 2001: 4). The economic agents who initiate the process cannot be allowed to exhibit memory, whereby the historical route through which the disturbance arose is rendered superfluous to the knowledge of how to act in the present (Maas 2014: 100).

The trick in operation here is the creation of determinacy out of a wholly indeterminate set of actual practices. The determinacy of the world within the model arises from the assumption that there is always a maximum condition and it will always act as a behavioural attractor. But we know from the indeterminacy of the world beyond the model that the economy does not display repeated starting conditions as every new development is historically unique. The essential condition of indeterminacy is thus being "corrected" by the assertion of an equilibrium state, but only through adopting "'the clipping and pruning' that Marshall warned us not to do" (Rugina 2005: 236). Once again, a rather large gap emerges between the second M and the second S in Samuelson's favoured SMMS narrative of disciplinary development. He appears to have reoriented economics away from truth-content propositions (Hausman 2008: 128). In place of what he considered to be such old-fashioned concerns he substituted the thoroughly modern logic of truth-form propositions. Whether he was thinking in these terms or not, this is consistent with the move to explicitly post-Cauchian metamathematical commitments. In this way, what can be treated as true is simply that which follows from the initial statement of conditions, even if those conditions belong to a purely hypothetical realm.

Samuelson (1972: 253) offered his equilibrium oriented mathematical structure as if it was all that economists needed to know, all that was necessary to reveal the most basic essence of modern economic life. However, in the same year he began his undergraduate studies at Chicago, Kurt Gödel published his famous incompleteness theorems. They stated that no consistent system of mathematical axioms is ever capable of proving its own validity (see Chapter 2). There must always be at least some undecidable propositions to allow them to work in their own terms (Smullyan 1992: 106; Shankar 1994: 24). That is, we have to think that Samuelson's mathematics works only imperfectly as statements of purely mathematical relationships, and this is before we consider the much more controversial assertion that the same mathematical structure is directly translatable into empirically meaningful insights into everyday economic life. If the mathematics is susceptible to objections as *mathematics*, just imagine how many more concerns are justified when the mathematics is presented as if it were synonymous with all that economists might achieve.

Gödel's incompleteness theorems reduce Samuelson's mathematically oriented behavioural axioms to a state in which they can be neither proved nor disproved as standalone objects of reasoning (Hodgson 2013: 81). The mathematics of constrained maximization must contain some other mathematical information for a solution to be derived, and in these circumstances it is impossible to tell whether apparently disconfirmatory evidence is challenging a hypothesis related to the maximization condition or a hypothesis related to the accompanying mathematical information. Nor, indeed, is it ever possible to know in any definitive sense whether the accompanying mathematical information acts as some sort of masking agent to prevent a direct refutation from taking place at all. This takes us into the territory of the Duhem–Quine

FALSE PROPHETS OF ECONOMICS IMPERIALISM

thesis, whereby no single statement, such as those associated with the economics of maximizing behaviour, is independently vulnerable to refutation. What might be used as evidence, after all, contains the imprint of assumptions that cannot be observed in isolation from the hypothesis to which they give meaning (Mandik & Weisberg 2008: 236). Even in their most simplistic form, Samuelson's market models contain too many situation-defining and derivation-facilitating assumptions to know for sure what might be tested when statements about the model world are translated into statements about the real world.

CONCLUSION

Samuelson's approach to mathematizing the market model has left many unanswered questions when viewed from literatures that are under threat of colonization by economics imperialists. But this should not lead us to overlook his accomplishments. Samuelson is surely at the top of the list of those who have done most to provide their fellow economists with their learned intuitions. According to his friend and fellow Nobel laureate Robert Solow, "If you did a time and motion study of what any modern economist does at work, you would find that an enormous proportion of standard mental devices trace back to Paul Samuelson's long lifetime of research" (cited in Frost 2009). The same could be said when trying to understand how economists' mathematical market models have helped propel the urge towards economics imperialism. *Foundations* turned constrained maximization problems into the benchmark against which economists evaluate the worthiness of competing interventions into debates. That benchmark has now secured a bridgehead in many other social sciences.

Samuelson was always fully aware of what mattered most to him: it was that he influenced in the most fundamental way how other people chose to see the world, not whether they obediently followed every step of his own chosen mathematical applications. He readily admitted that "by the time *Foundations* celebrated its official twentieth birthday [in 1967], its pages of ... calculus were old hat" (Samuelson 1983a: xviii). Yet this did not stop him from using his Nobel Lecture three years later to continue championing the maximization principle as the best possible lens for visualizing the basic essence of eco-nomic problems (Samuelson 1972: 256). Perhaps still today Samuelson is the key influence, not only on how other economists view the world, but also on the fact that the world they spend most time looking at is that within various market models. Significantly, the production of outcomes in the model world follows the adoption of a particular mathematical logic, not the accumulation of relevant economic facts.

Samuelson almost certainly left his most profound imprint on economics by showing how mechanical analogy might give way to mathematical analogy (Ingrao & Israel 1990: 182). He had a habit in the twilight of his career of pointing in a mock self-deprecating manner to the feeling that his "Newtonian calculus" – indeed, sometimes his appropriation of "Cournot's Newtonian calculus" – made him a hostage to methods that had fallen from fashion (Samuelson 1983a: xviii, xvii). The implication was that a younger Samuelson who was coming to age intellectually today would have embraced mathemati-cal methods that were much more obviously *au courant*. However, we should not be misled, because there was nothing that looked particularly Newtonian about his methodological choices when undertaking the transition from mechanical to mathematical analogy. Isaac Newton achieved great fame in the 1680s for unifying previously separate explanations of celestial and earthly motion in a single system of equations (see Chapter 1), and for a unificationist like Samuelson any association with Newton's name was clearly attractive. But within Newton's general argument pattern each physical quantity had to be calculated directly through observation (Grosholz 2016: 67). Nothing of this nature is apparent in Samuelson's work, despite the superficially deferential appeals to operationally meaningful theorems.

Hilbert and Bernays (2003 [1934]: 2) cited Newtonian mechanics approv-ingly as concrete axiomatization based on inhaltliche statements of substan-tive content. Here the important distinction is that Newtonian calculus was a branch of algebra, but Lagrangian calculus was a branch of analysis (Rao 2011: 49). For Lagrange, physical explanation could be derived directly as a matter of mathematical logic, without the bother of having to go out into the world to discover what was happening beyond the model (see Chapter 1). This looks much better as a description of Samuelson's work, and why would that come as a surprise when he specifically mentions his introduction to Lagrangian techniques as the moment he first opened his mind to the full range of possibilities inherent in mathematical applications (Backhouse 2017: 67). In seeking to demonstrate how mathematical economics might follow the Lagrangian path to also become a branch of analysis, Samuelson authorized an approach in which explanation could proceed in the absence of even a deferential nod towards calculability. This seems to place economists' market models beyond the realisticness safeguards of Cauchian metamathematics.

The advancements that Samuelson announced for economics have equally helped advance the agenda for economics imperialism. Adopting his mindset is to consciously expand the types of experiences that can be studied using market models. Samuelson (1981: 8) was the arch-separatist, even more so in his own mind than Robbins (see Chapter 5). He was determined to let nothing beyond commonly accepted principles of economic theory into his market models except for the mathematical techniques that brought them to life. Any aspect of economic experience that does not conform simply to the ideal of perfect market relations is unrecognisable as a suitable research topic, but the precision in mathematical expression in *Foundations* also enhanced how much of social reality could be thought about as if it inhabited just such a pristine market world. All social institutions, as well as all social relations that took shape within those institutions, could henceforth be modelled on experiences that might be nothing more than an historical anomaly in the world beyond the model. Solow again on Samuelson: "he had a marvelous intuition about how a market economy had to be. 'It must work like this', he would say. 'Now all we have to do is prove it'" (cited in Frost 2009). The "we" in question has moved progressively outwards in Samuelson's wake from economic theorists to economics imperialists.

However, Samuelson's economic agents can only exist in such historically decontextualized circumstances if the rationality they display for reasons of mathematical convenience exhibits a peculiarly motiveless form. These agents are maximizers through instinct, but they are not allowed to reflect even for one moment on why they should always make maximization their goal. They serve as convenient props for facilitating the boundary-hopping activities of economics imperialists, because they dissolve the need to engage with culturally embedded explanations of behaviour that appear in the subject specialist literatures they are seeking to displace. The ignorance of empirical debates to which many economics imperialists freely admit cannot be turned against them if detailed case histories of human behaviour are superfluous to the task of building the explanatory model. However, this works only if the overall objective is not to explain anything beyond the limits of the model world.

Samuelson's uncharacteristic modesty about the mathematical limits of his model worlds is acknowledgement that other, more mathematically gifted economists were pointing towards formalisms that were much more precise than his. They were making impressive headway in further breaking down barriers within the field, even as he was taking most of the plaudits for founding a new economics based on mathematical analogy. Samuelson's mathematics have always occupied a curious double place in the consciousness of his fellow economists: alarmingly complex for those with no mathematical training, but disarmingly simple for those with more than him. The next step in the prehistory of economics imperialism is to explore the defining moments of the formalist revolution of the 1950s. It poses problems for those economists who have sought to colonize other social science subject fields, because it is the only one of the four phases outlined in this book where the direction of travel is likely to unnerve them. Those who have pursued the formalist revolution with the greatest mathematical skill not only overrode all of Samuelson's mathematical structure but also cast serious doubt on the market template which he imposed on his constrained maximization problems. His approach long ago had to cede its position at the mathematical frontier of economic theory. More damningly, those who assumed that mantle also demonstrated quite conclusively that Samuelson's mathematical models failed important tests of internal rigour.

CHAPTER 7

Axiomatization as proto-imperialist move 4: Arrow–Debreu and the search for economically meaningful existence theorems

INTRODUCTION

Judging by their actions, economists generally seemed happy to be taken to where Paul Samuelson wanted them to go. In the years between him starting his Harvard PhD in 1935 and receiving his Nobel Prize in 1970, the content of economics journals increasingly came to prioritize reasoning through mathematical objects (Backhouse 1998b: 86). However, Samuelson cannot claim sole credit for such a shift, despite his own history of mathematical economics repeatedly stressing the blank canvas he initially faced (Samuelson 1986: 797). This might have been true of his immediate environs in Cambridge, Massachusetts, but by no means elsewhere. The Cowles Commission was simultaneously bringing together a collection of mathematically minded scholars in Chicago, but outside the University's Economics Department where Samuelson had enjoyed a less than wholly fulfilling time as an undergraduate. All were influenced either through direct participation or inherited intellectual objectives by Karl Menger's Mathematical Colloquium in pre-war Vienna (Leonard 2010: 154).

While Samuelson was always both a willing and an effective advocate for economists with an interest in mathematizing their field, a significant proportion of the Cowlesmen were mathematicians first and foremost, who had merely happened upon economics as a convenient focus for their writings (Mirowski 2012a: 141). What looked like a significant advance in mathematization from

Samuelson's perspective was often met with a shrug of indifference by the more mathematically sophisticated Cowlesmen. Samuelson might have been able to help them add more convincing economic narratives as counterparts to their mathematical reasoning, but that was all. From their perspective, he had produced a very basic mathematical framework for facilitating his economic arguments, whereas they were interested in determining the full logical implications of mathematical propositions, the economic relevance of which was at best of only secondary importance. The achievements that gave the Cowlesmen most satisfaction were consequently often fundamentally unreadable from a Samuelsonian perspective, so unforgiving were their mathematical demands on the reader. Samuelson, though, was reluctant to take his usurpation lying down. Looking back at the clash of styles in postwar mathematical economics, he used the 1983 Introduction to the enlarged version of *Foundations* to say, "More can be less. Much of mathematical economics in the 1950s gained in elegance over poor old Pareto ... But the fine garments sometimes achieved fit only by chopping off some real arms and legs ... Easy victories over a science's wrong opponents are hollow victories" (Samuelson 1983a: xix).

Despite his protestations, it is generally agreed that the cutting edge of postwar mathematical economics was only very temporarily associated with the Samuelsonian tradition. This does not make him any less important to the prehistory of economics imperialism, where his influence continues to cast a long shadow, but his contribution to economic theory needs to be placed in the correct perspective. In the increasingly esoteric realm into which mathematical economics was propelled at the height of his career, Samuelson was hardly ever more than an interested bystander. His main role appears to have been to prise the door ajar, after which it was left to more accomplished mathematicians to reconstruct the field. Samuelson's revealed preference theorem asserts the behavioural content of economic agency prior to any actual empirical study having to occur, and this licenced the formalist revolution of the 1950s (Hands 2009a: 160). For the practitioners of formalism, the mathematical form of the argument takes complete precedence over its economic content; proof-making in the world within the model thus fully exhausts their attention.

The most important existence proof for mathematical economics was provided by Kenneth Arrow and Gérard Debreu in 1954. It states the conditions under which a position of general economic equilibrium can be proved, where those conditions are the situation-defining and derivation-facilitating assumptions that provide the system of equations with an imagined economic character. Samuelson had implored economists to require their subject field to revolve around ever more elaborate mathematical representations of the economy (Rugina 2005: 237). His own work, however, completely avoided the question of how the solution to a set of general equilibrium equations might be demonstrated (Weintraub 1985: 85). *Foundations* was a hangover of what the Cowles Commission theorists saw as a long-bypassed tradition. It operated on the basis that if it could be shown that the number of unknowns in the equations was the same as the number of equations, then it could be said by the laws of elementary algebra that it was certain the mathematical system must have a solution (Samuelson 1983c: 839). He had toyed with the conception of mathematics as a branch of analysis through his embrace of Lagrange multipliers, but he had never gone all-in with the implied break with algebraic objects. In his anti-analysis mindset, there was no further requirement to state what the solution actually was, because it was considered sufficient to use the equation-counting approach to know that the solution must be out there somewhere.

However, nobody associated with Menger's Mathematical Colloquium ever took the equation-counting approach seriously (Debreu 1984: 268). They argued instead for "the actual demonstration of whether or not such systems have a solution" (Morgenstern 1951: 363), implying that mathematics was a branch of analysis or it was nothing at all. Oskar Morgenstern (1941: 370) had written a scathing review of John Hicks's *Value and Capital*, accusing him of being "systematically incorrect" in his view that the determinateness of a system of equations could be demonstrated merely by having matching numbers of unknowns and equations. Meanwhile, Morgenstern's collaborator, John von Neumann, the other great inspiration underpinning Cowles Commission economics, refused the invitation to review Samuelson's *Foundations* that was written in the Hicksian mould on the grounds that, "He is no mathematician". "You know, Oskar", von Neumann wrote to his friend, "if those books are unearthed sometime a few hundred years hence, people will not believe they were written in our time. Rather they will think that they are about contemporary with Newton, so primitive is their mathematics" (cited in Israel & Gasca 2009: 132).

The mathematical economics profession quickly paid homage to the advances contained within Arrow and Debreu's paper. That sense of respect is no less pronounced today. The 1954 *Econometrica* paper is, after all, the only one in economics to have contributed directly to the award of two Nobel prizes: Arrow in 1972, Debreu in 1983. Arrow and Debreu are still treated "with awe, and not a little apprehension" (Weintraub & Mirowski 1994: 256). The word "classic" is a favourite descriptor of various elements of their published research (Varian 1984: 10) and, as with other "canonical work" in the subject field (Currie & Steedman 1990: 129), it became firmly established as part of the "initiation programme" for economics PhD students (Beed & Kane 1992: 603). Neither Arrow nor Debreu in retrospect seems to have wanted to recognize their existence proof as being amongst their best work (Horn 2009: 79). But this has made little impression upon the reverence that typically underpins collective disciplinary memory. It is still consistently presented as the ultimate "frontier of basic research" (Roncaglia 2005: 348), as having "set the postwar standard" for others to emulate (Davis 2013: 90), as the "foundation" on which formal mathematical proof-making in economics is built (Black, Hashimzade & Myles 2012: 16), and as the "benchmark" for everything that followed (Backhouse 1994: 219; Brock & Colander 2005: 31). But a benchmark for what, precisely (Bruno 2010: 56)?

As Alan Coddington (1975: 544) noted 50 years ago, general equilibrium constructions have tended to take on a life of their own following the publication of Arrow and Debreu's existence theorem. The questions that are asked about the world beyond the model are all too frequently led by what the theory can say about the self-made world within the model. At most, then, Arrow and Debreu's famous paper might serve as "a useful benchmark of idealization", a series of statements couched in formal mathematical logic that depict a purely hypothetical system far removed from any known state of affairs (Maskin 2001: 51). Their existence theorem might well encapsulate "the home of efficiency", but this does not mean that the actual institutions of the economy can ever perform a similar function (Gale 1982: 198). The most compelling result of the general equilibrium tradition might therefore be to recognize how far mathematical economics has come simply to cast doubt on its own economic achievements. This would obviously make uncomfortable reading for economics imperialists, but that does not make it less true.

In an attempt to pursue this line of argument and to show how axiomatization acts of the final proto-imperialist move in my story, the chapter now proceeds in four stages. In section one, I ask about the *economic* problem to which Arrow and Debreu's existence theorem is offered as a solution. Sections two and three shed additional light on how their thinking was shaped before coming together for the 1954 *Econometrica* paper. In section two, I trace the influence of the logician Alfred Tarski on Arrow and, in section three, the influence of the mathematical collective Nicolas Bourbaki on Debreu. Arrow remained true throughout his career to the objective of seeking practical applications of a clearly interpretable economic nature for his mathematical proofs, whereas Debreu took the hardline approach that the proofs provided solutions to purely mathematical problems that stood independently of any subsequently inferred economic meaning. Yet both asserted the primacy of the axiomatic method, whereby formal deduction from basic propositions allows an entire mathematical system to be created in its own terms. It was other people who subsequently treated their mathematically oriented solutions as established and even self-evident economic facts. In section four, however, I show that those facts have increasingly come undone. Culminating in the now famous Sonnenschein–Mantel–Debreu impossibility theorems, this more recent work in the Arrow–Debreu tradition reveals just how little is left standing of orthodox models of market exchange following ultra-rigorous mathematical treatment. General equilibrium economics post-Sonnenschein–Mantel–Debreu works most effectively to draw attention to its own highly restrictive conditions of application. The fourth phase in the prehistory of economics imperialism might therefore turn out to be the most important, but only in a negative sense. It compels us to ask whether the previous 100 years of intellectual endeavour have led only to a dead end.

THE ADVANCES OF ARROW–DEBREU

The 1954 *Econometrica* paper quickly passed from presenting a pretty much unreadable solution to the existence of general economic equilibrium to being part of the subject field's folklore. It proved exceptionally difficult to find suitable referees, with one of the journal's editors being forced into acting as a *de facto* third referee to break the tie between a mathematician who doubted whether the equations contained any useable economic knowledge and an economist who doubted his ability to follow the logical steps required to demonstrate the solution (Weintraub & Gayer 2001: 440). The economist's wish to see the paper published eventually won out, and mathematical economics was raised to ever more esoteric heights on the back of results that almost no other economist at the time could verify. It is comforting to think that a proof has been replicated by countless people following each of its formal steps in identical sequence, but nothing of that nature can be claimed in this instance. It serves as the most obvious example in economics of what philosophers of mathematics call proof by authority (Hoffman 1998: 200). Arrow and Debreu were trusted by their peers to have got the mathematics right, even if almost nobody could say for sure that they had.

The question remains, though, of what, exactly, the Arrow–Debreu existence proof is a proof of. Arrow and Debreu's axiomatic refounding of the Walrasian tradition allowed them to explore through logical deduction alone the full implications of starting with a series of basic propositions that brought a solution within their reach. But as for what this genuinely proved, it might have been nothing more than the need to assert the presence of a continuous market system for all future conceivable states of the world within the model if a similarly continuous market system was to be conjured in the present. Even the situation-defining assumption of perfect competition is not enough on its own to allow the general equilibrium equations to elicit a solution in the model world if prices cannot be ascribed to everything that might possibly be traded in some future state. Yet this in itself can count as evidence of an advance. The search for a demonstrable existence theorem was the prize that had eluded economists for almost a century since Léon Walras's *Éléments d'Économie Politique Pure*. Arrow (1987: 198), for instance, has justified all the effort that has been expended on the existence theorem because it has required economists to be more rigorous in their specification of economic terms if they are to satisfy the demands of their most mathematically proficient colleagues.

These developments provided additional impetus to Samuelson's (1981: 4, 1983b: 792, 1986: 797) attempts to rid economics of utility theory, because they showed that it was unnecessary to go beyond the stipulation of demand and supply functions to populate the general equilibrium equations with well-behaved economic agents. The integrability literature successfully rationalized all individual economic behaviour by showing as a matter of mathematical logic that every activity on either side of the market could have been produced by a budget-constrained individual acting solely upon a max-imization principle (e.g., Uzawa 1960; Hurwicz & Uzawa 1971; Hurwicz & Richter 1979). This has the effect of stabilizing Walras's demand and supply functions, rendering them susceptible to axiomatic treatment (Hands 2012: 393). Under such a system, the overall objective is simply to discover the impli-cations of starting with one set of mathematically mouldable propositions and not another. "Simply" is maybe the wrong word when looking at the pages of often forbidding mathematics, but the task is still no more than to ascertain through logical deduction what a well-ordered mathematical system looks like on the basis of its starting propositions. In the process of restricting the analysis in this way, though, Lionelli Punzo (1991: 7) suggests that it changed the discussion from Walras's intuitionist account of everyday equilibration *processes* to an avowedly formalist equilibrium *model*.

Yet still, ironically, many of the analytical weaknesses of the more substan-tive features of Walras's account remained unresolved. The mathematics, after all, cannot do everything, however sophisticated it is. There is no explanation in the 1954 *Econometrica* paper for how the economy actually manages to reach the desired state of equilibrium. Indeed, Arrow and Debreu (1954: 266) made it clear that this was a question they were willing to sidestep for expos-itory purposes, and it is a question that Debreu, for one, continued to avoid for the rest of his career (Hildenbrand 1983: 26). It is doubtful, then, whether the modern general equilibrium tradition has gone much beyond Walras in creating an approach that works at any level other than a purely theoretical account of the world within the model (Händler 1980: 51).

Walras attempted in the *Éléments* to explain market self-equilibration on the basis of equations borrowed from physics (Beinhocker 2006: 30). His first efforts involved imposing a *tâtonnement* process he believed captured rea-sonably accurately real-world events (Jaffé 1967: 12). There is no direct trans-lation of "*tâtonnement*" into English, but it is used to highlight the sense of an equilibrium-oriented system seeking a position of rest and "groping" its way towards it. The image of a stationary state is consistent with the mathematics of turning points captured by the differential calculus, while the image of incrementally approaching a more-or-less stable condition invokes at least a quasi-economic process. However, the precise character of this process has been the source of so much of the criticism of Walras's system. "[T]imeless tâtonnements in a barter economy" is how Roy Weintraub (1977: 2) describes its clearly hypothetical economic content: not really what we would instinctively think of as a market at all. Arrow and Debreu's core continuity assumptions still hint very strongly of a residual timelessness, as does the imposition of an economic structure of fully complete futures markets. Their project was no improvement on Walras's when it came to providing a genuinely economic explanation of the equilibration process (Morishima 1984: 60). "Nirvana might be there", writes Daniel Fusfeld (2002: 217), "but how does one get to it?".

The *tâtonnement* concept has only ever been used because it fits the story of the model world that the mathematics is able to tell, not because it helps to describe the structure actually displayed by the world beyond the model. *Tâtonnement* does not work even in its own terms without the introduction of an auctioneer construct, which is no more persuasive as an economic explicatory device when dressed in the mathematical *haute couture* of Arrow and Debreu than when wearing the more ascetic mathematical garments of Walras. At most, the auctioneer construct maybe changes from a situation-defining assumption for Walras to a derivation-facilitating assumption for Arrow and Debreu, but either way it appears to be a contrived solution for the absence of genuine economic explanation. Arrow's (cited in Weintraub 1985: 104) "discovery" that his predecessors in the general equilibrium tradition had really been using a "disguised fixed-point argument" all along did nothing to put more economic flesh on the *tâtonnement* process when he used an actual fixed-point theorem to insert Nash games into Abraham Wald's proto-existence proofs. Issues of mathematical tractability were obviously driving the theoretical endeavour in a way that could not ensure the presence of only non-trivial economic interpretations of general equilibrium (Ackerman 2002: 61). This might have rendered the system of equations more stable, but it could not do likewise to the Walrasian *tâtonnements* that the equations were designed to describe.

Walras had anyway taken the decision in the fourth edition of the *Éléments* to use a "pledges" model to allow the equilibrium state to be arrived at instantaneously rather than, as before, through leaving behind various disequilibrium states (Jolink 1996: 27). Perhaps, then, the auctioneer construct was a derivation-facilitating assumption all along, certainly in the context of the pledges model. For no reason other than that his mathematics demanded it, Walras had replaced an intuition about iterative trial-and-error pricing structures that did not look too far removed from observable economic reality, ending up with an explanation of market-clearing dynamics that works simply by definitional fiat (Walker 1987: 767). Arrow and Debreu's existence theorem looks equally suspect on these grounds, because they employed an almost identical sleight-of-hand to disqualify any discussion of progress through disequilibrium before eventual arrival at the promised land.

Debreu had been uneasy about relying on an auctioneer construct from his earliest introduction to Arrow's work on the existence proof. In the very first piece of correspondence he sent to his future writing partner – both were working for the Cowles Commission by 1952, but had yet to meet in person – he made the following comment on Arrow's initial forays. "The introduction of the fictitious players I + 1 $\leqq j \leqq$ 2I with the use of Kuhn and Tucker's theorem seems artificial to me ... [T]his is probably my most important criticism" (cited in Düppe 2012b: 498). Yet two years into their joint venture when the existence proof was ready for publication, there was still "a fictitious participant who chooses prices, and who may be termed the *market participant*" (Arrow & Debreu 1954: 274, emphasis in original). The problem with the auctioneer construct, of course, is that it removes all of the most obviously economic content from the explanation of the equilibration process (Dûppe 2011: 73). All the leading lights of the general equilibrium tradition have expressed concern about what its presence implies for the underlying conception of agential decision-making (Arrow & Hahn 1971: 34). Imposing axiomatic structure on all decision-making functions is fine if the demonstration of equilibrium is simply a mathematical exercise. However, when a global "Leviathan" has to be assumed into existence to capture the sense of an operative outcome-oriented price mechanism (Clark 1992: 166), a specifically *economic* equilibrium can only be reached "by decree" (Ingrao & Israel 1990: 331).

This is not the most obvious image of a functioning market, though, so much as its polar opposite of a centrally planned economy. But perhaps it is unsurprising that comparisons have been drawn between Arrow and Debreu's market participant and an ideal-typical central planner (Shubik 1977: 215). After all, economists have struggled since the 1920s to formally differentiate markets from planning under neoclassical treatment (Lavoie 1985: 27; Bockman 2011: 46). The 1954 *Econometrica* paper therefore merely reflects its broader professional origins in leaving unspecified what sort of economic institutions it might be taken to refer to (Borglin 2004: 26). The language of markets is ever present throughout the paper (Arrow & Debreu 1954: 265, 279, 271, 272, 274, 275, 278, 287), but the formal representations around which that language appears do not relate exclusively to markets.

The mathematical developments enacted on the back of the Arrow–Debreu existence proof render everyone within the model world entirely passive, in a manner completely unknown within actual systems of decentralized markets (Cialowicz & Malawski 2011: 34). What results, in Jon Mulberg's (1995: 71) telling phrase, is "dictatorship over economic choice". There is a pervasive anti-individualism that runs right to the heart of general equilibrium economics, to the point at which its most essential building block, the existence theorem that provides the benchmark for all that follows, eliminates all agential heterogeneity (Sent 1999: 730). People cannot ever be allowed to navigate their preferred way through the economic world by making their own minds up when placed in a choice situation. As we move through the second, third and fourth phases of the prehistory of economics imperialism, we see how Robbins's economizers have become increasingly overpowered as conscious agents by the mathematical objects in which they have been situated. They have been reduced to the purely functional alternative that best serves the purpose of the mathematical proof. Given the continuity assumptions being deployed by Arrow and Debreu, the auctioneer construct always acts by proxy on behalf of every conceivable person, both now and in the future (Currie & Steedman 1990: 137). People can have preferences ascribed to them, but they cannot be permitted to experience the sensation of *having* a preference (Davis 2013: 92). As Robert Clower (1995: 314) has noted, such agents "appear to be placeholders for 'plans' or 'thoughts' fathered by the wishes of a single 'principal' ... the 'theorist'".

Model worlds following in the Arrow–Debreu tradition therefore look wholly at odds with claims so frequently made in their name. Somewhere deep inside the fundamental lemmas of general equilibrium theory, we are told, lies the answer to the centuries-old question of how Adam Smith's invisible hand is activated economically (Nadeau 2003: 56). So many appraisals of the general equilibrium tradition settle on the idea of a largely uninterrupted line of scholarship that runs from Smith via Walras to Arrow and Debreu (the seminal account is in Schumpeter 1984 [1954]: 189). This is an alternative but equally historiographically suspect timeline to Samuelson's favoured SMMS approach that places him at the end of a list of greats that also includes Smith, Mill and Marshall (see Chapter 6). Arrow and Debreu's *Econometrica* paper is presented as having achieved the formal treatment which lent real analytical purpose to Smith's initial creative insight (Black, Hashimzade & Myles 2012: 15). However, this is what it singularly fails to do. The mathematics obscures almost everything of economic significance that might be read into the invisible hand metaphor. Was what Arrow and Debreu set out to do therefore ever really economics at all?

ARROW AND THE MATHEMATICAL LOGICIAN'S ROUTE TO THE EXISTENCE THEOREM

Mathematical economics stood at a crucial juncture in the fourth phase of the prehistory of economics imperialism. For some, the task was to solve well-defined economic problems using increasingly complex mathematical reasoning; for others, it was to demonstrate mathematical virtuosity and then look for economic interpretations in the resulting lemmas. Mathematical economists were thus arrayed across the spectrum of scientific unification, even if the vast majority were much closer to the derivational unification pole than the ontological unification pole. Even the seminal paper on the existence of general equilibrium was written as a compromise about just how much priority should be given to presenting the mathematical proofs solely in their own terms (Düppe 2012b: 492). Arrow (1987: 195) – with his reputation as the "economist's economist's economist" (Niehans 1990: 497) – understood his role as making the existence theorem as accessible as possible to his peers. Yet Debreu (1987: 249) – the "mathematical economist's mathematical economist" (Samuelson 1983c: 838) – seems to have taken more personal pride in the purely mathematical spin-off piece he published alongside, and as a potential alternative to, the joint paper (see Debreu 1952).

Ingrao and Israel's (1990: 257) path-breaking work on the history of the existence theorem suggests that there are four distinct foundation stones for Arrow and Debreu's 1954 *Econometrica* paper. It is best viewed, then, as a forced convergence of distinct themes (Weintraub & Gayer 2001: 421) through a process that Marcel Boumans (1999: 90) calls "mathematical moulding". Two of the four facilitating research agendas related to the mathematical ground-work for the existence proof. John von Neumann had pioneered the use of convexity techniques and fixed-point theorems (Dimand & Dimand 2002: 19). Via the interest that members of Menger's Viennese Colloquium had shown in their economic implications, von Neumann's original insights were fleshed out further in four key papers written by Abraham Wald (Becchio 2009: 2). Wald had escaped from Europe in the 1930s and was employed at Columbia University when Arrow enrolled there to write his PhD under the supervision of Harold Hotelling. Wald was appointed as Arrow's advisor, but he sought to steer him away from using his doctoral studies to focus on the existence of gen-eral equilibrium, thinking that as much progress as was possible had already been made on the mathematical front (Arrow 2002: 1). Arrow's initial interest in the existence theorem therefore seems to have been triggered by the two more obviously economic developments that shaped the 1954 *Econometrica* paper. These were the early Samuelson's attempts to reposition economics as the study of maximizing conditions and the challenge to the significance of immediate economic realism of which this was merely one example (Ingrao & Israel 1990: 257).

There is an important dating issue to be taken into account here. Attention turned to exploring general equilibrium theory mathematically for the first time in the 1920s when von Neumann began to develop tools suited to the task (Debreu 1986: 1265). However, attention turned *back* to exploring general equilibrium theory economically only in the 1930s (Backhouse 2002: 254). The renewed interest in the economic dimension was sparked by John Hicks, who brought the Lausanne School tradition of Walras and Pareto more directly to the attention of English-speaking economists (Weintraub 1985: 83). As the previous chapter showed, Samuelson also followed in Hicks's footsteps. He picked up on Hicks's advances in consumer theory to argue that every prob-lem of interest to the economist might be established as a problem of maxi-mization within the context of a one-market partial equilibrium framework (Samuelson 1941: 97). Arrow (cited in Horn 2009: 72) also admitted to having been "excited" by this aspect of Hicks's work. His interest, though, was how it might be possible to appropriate the extra rigour Hicks had given to Walrasian economics in an attempt to scale up Samuelson's maximization problems and generalize to the economy as a whole (Arrow 1987: 203).

After having taught himself Hicks's system through close textual study of *Value and Capital* as an undergraduate, Arrow began to wonder what might be learnt if he went ahead and solved Hicks's equations for him. "I guess I had been exposed to enough mathematics", he later reflected, "to know that when one has a system of equations one worries about existence" (Arrow 1987: 194). When Hicks visited Columbia in 1946 to deliver a lecture, the PhD student Arrow challenged him in public on a point where he knew he was already in possession of mathematics that would help him push beyond *Value and Capital* (Arrow 1983: 2). Hicks might have breathed new life into the economics of general equilibrium, but the future research programme envisioned by Arrow required further *mathematical* advances of the underlying market model, not economic.

Hicks actually had two general equilibrium models: one in which the transaction of all conceivable goods in all conceivable states of the world takes place at a single point in time, the other in which a market is made each "Monday" for all transactions taking place in that "week", but only then (McKenzie 1989: 11). The one that appealed most immediately to Arrow had dated goods that would change hands between consumers with perfect foresight only when the characteristic associated with a particular date exactly met the individual's preference (Mandler 1999: 40). He was therefore eager to retain an element of economic realisticness in his proof-making, as long as the mathematics would permit it. As the market for each good was reinvented from scratch on each new date when more supplies became available, consumers with perfect foresight could stretch out entry and exit decisions indefinitely. Such dynamics could receive formal representations in the continuity assumptions of the mathematics of convexity, while still allowing the general equilibrium system to be built upon the established economics of consumer choice (Gass & Assad 2005: 70). The use of a fixed-point theorem to solve the system of equations was not yet on the horizon, so a gap still had to be bridged with the von Neumann–Wald route to the 1954 *Econometrica* paper.

Arrow remained wedded throughout his career to a philosophical instrumentalism, whereby to retain relevance as an economist it was always necessary to show that economics could be useful (Beed & Kane 1992: 601). The focus on immediate real-world implications – however abstractly he thought about their manifestation – was always his guide for fitting together the separate pieces of the general equilibrium jigsaw. The entire existence proof was of no value in itself, only insofar as it could open up new windows on the world (Düppe 2012b: 508). General equilibrium theory "just says 'go out and calculate'", he argued (Arrow 1987: 205). Arrow refused to embrace Samuelsonian "directness" in his proof-making if the additional mathematical precision took him further away from being able to frame political debates about the desired form of everyday economic institutions (Arrow 1987: 201). Not for nothing did he insist that his primary economic interest was in understanding "the development of economic planning" (Arrow 1983: vii). Lurking in the background once again is the formal identity of neoclassical models of competitive and centrally planned economies (Bruno 2010: 56).

This clear commitment to the normative implications of their existence theorem raised tensions in his working relationship with Debreu. This was not, it has to be said, that Debreu had any objection to the particular political content that Arrow wanted to read into the general equilibrium approach, only that he thought no such content should be admitted (Debreu 1986: 1266). During the process of corresponding about formative drafts of what became the 1954 *Econometrica* paper, Debreu (cited in Düppe 2012b: 503) warned Arrow against "forced interpretations of ancient texts" that might present the existence theorem in anything other than its own pure mathematical form. So much, then, for the widely held view that Arrow and Debreu had worked tirelessly to prove the intrinsic economic meaning of Smith's invisible hand metaphor. The sense that the writing partners were inadvertently engaged in two entirely separate intellectual projects was nowhere clearer than in Debreu's suggestion that all references to the normative implications of their existence theorem should be removed to allow the mathematics sole priority. Notwithstanding Arrow's (1987: 194) insistence that the eventual joint publication owed more to Debreu's starting position than to his, the final version still describes their aim as being "of interest both for descriptive and for normative economics" (Arrow & Debreu 1954: 265). Arrow was prepared to sacrifice a degree of mathematical generality in the final version of the existence proof so that it might have more practical purchase in relation to the real world (Düppe 2012b: 500). Yet the axiomatics were nonetheless never very far away, and the method of logical deduction from mathematically tractable assumptions remained paramount.

A tension exists in Arrow's work regarding the relationship between the world in his mathematical market models and the world beyond. On the one hand, he has an evident commitment to harnessing economic knowledge to social programmes of reform. The spirit of Jevons thus appears to linger. On the other hand, he placed the production of economic knowledge within the context of the Quinian philosophy of science that proved increasingly influential within postwar American intellectual circles (Davis 2013: 92). Subjective propositions had no place within scientific enquiry, according to the philosopher W. V. O. Quine, the exact opposite of Jevons using his attack on establishment economic thinking as a constitutive element of his theory. Yet the precision that followed from eliminating subjective propositions was to be put to use, in Arrow's hands at least, to create clear normative statements about the outer limits of market self-regulation. Consequently, there are two distinct moments in Arrow's economics that are reminiscent of Robbins's ultimately unconvincing dualism of economic science and political economy (see Chapter 5). One occurs at the point of discovery, which conforms to the most basic Quinian commands to produce work that is incontrovertibly true in its own terms. The other occurs at the point of use, which takes its most immediate cues from Arrow's own conscience. This separation may be no more convincing than Robbins's.

Arrow's Quinian world originates in the work of the Polish logician, Alfred Tarski. Quine (1960: 27) was himself influenced by Tarski to develop a purely formal semantics for science, one in which there could never be reasonable doubt – at least not amongst the initiated language users – as to what was being said. Meaning was thereby reduced to relations cast in set-theoretical terms between names and predicates, so that a single representation is all that is possible for any given object (for instance, one described by the name "the economy") or for any given property of a system (say, one described by the predicate "equilibrium"). Arrow permitted disagreement over how the economy might be organized and over whether a particular equilibrium was desirable, but not over what the economy and equilibrium were in themselves. Hence, there are clearly demarcated boundaries between social and scientific facts. The Quinian-Tarskian ideal of social facts relates to where the individual places the outer limits of the conscionable, whereas the Quinian-Tarskian ideal of scientific facts rests on "the realist bivalence assumption of just two truth values, namely the true and the false" (Boylan & O'Gorman 2008: 21).

Arrow had a front row seat when Tarski's ideas first began to leave their mark in the US. He had read much of Bertrand Russell's work on logic and was due to receive instruction at City College of New York from the visiting Russell in academic year 1941–42, only for the authorities to refuse his permit. The last-minute substitute was Tarski, who in 1939 had been in the United States at a conference when war broke out in Europe, preventing him from returning home. Somewhat incongruously, then, a renowned global star in the field of logic found himself teaching students at a local college. Arrow took full advantage of his good fortune to sign up for Tarski's course on the calculus of relations. "[W]hat I learned from him", he recalled, "played a role in my own later work – not so much the particular theorems but the language of relations was immediately applicable to economics. I could express my problems in these terms" (Arrow, cited in Feferman & Feferman 2004: 134).

The particular appeal was Tarski's "axiomatic treatment of relations" (Arrow, cited in Kelly 1987: 44). Here, establishing truth through mathematical axioms as if in a Hilbertian world allowed Arrow to push consumer theory beyond the Samuelsonian conventional truth of revealed preference to a Quinian literal truth. This was not to deny the good sense of Samuelson's approach as an economic starting point – Arrow repeatedly adopted it for his own work – but Samuelson's mathematical economics plus Tarski's mathematical logic provided greater scope for formal precision than Samuelson alone. Tarski's formalization of the calculus of relations provided an equally formal semantics of consumer behaviour, something that eluded Samuelson even as he hinted that it would be desirable. Consistent with the major claims of Hilbertian metamathematics, the use of Tarski's symbolic logic allowed Arrow (cited in Horn 2009: 73) to redescribe the economic concept of preferences as the purely logical concept of orderings. He thus got to treat it as indisputably true in its own terms. Robbins's concern to turn economics into a science of choice (see Chapter 5) thereby re-emerges, but now embedded in highly sophisticated mathematical reasoning tools.

Tarski had also left an impression on those who preceded Arrow in investigating how an existence proof might be derived to formalize Walras's intuitionist understanding of general equilibrium. In 1937, he was invited to write a piece called "Mathematics and Logic" for the in-house journal of Menger's Viennese Colloquium (Hands 2001: 73). He found himself lecturing to a group that had recently discussed in great detail Wald's early attempts to provide a mathematically robust existence theorem, and Tarski's analysis of the mathematics of formally deductive systems sparked fresh interest in this regard

(Becchio 2009: 18). He showed the assembled company how it might be possible to redefine all matters of reference solely in terms of mathematical logic (Halpin 2013: 98). In future, then, what might matter more was not how the equations of general equilibrium could be described using ordinary language techniques, but how they could be defined under a purely axiomatic treatment that rendered all matters of economic meaning unimportant. Tarski, it should be noted, never discussed scientific truth through concern with how it promoted meaning (Glock 2003: 103). We see here the potential for all pretence to finally be dropped that the mathematical objects of formal economic proof-making could be anything but defined functions. We also see the potential for a conclusive break with even a deferential nod to lived economic experiences when adjudicating on truth claims relating to the world within the model. Once the existence of general equilibrium could be reconfigured in these terms, the economics of the approach could shift from the realm of the mathematical logician to that of the mathematical purist. Enter Debreu.

DEBREU AND THE MATHEMATICAL PURIST'S ROUTE TO THE EXISTENCE THEOREM

The reaction to Debreu's Nobel Prize tells us much about his standing within the profession. Mathematicians celebrated the award as if he was one of their own (Basile & Li Calzi 2004: 112). Moreover, the interviews he was required to give to mark his success also made it clear that, had it been down to him, he would have been accepting the award for his achievements in mathematics, not economics (Düppe 2012a: 439). Debreu's embrace of economic theory in general and of the Walrasian tradition more specifically looks to be merely incidental to his wider intellectual aspirations (Punzo 1991: 3); the equations of general economic equilibrium captured his imagination simply because he assumed they had to be mathematically solvable (Warsh 1993: 86). Debreu was the puzzle-solver *par excellence* within economics, consistently throwing himself into research for its own sake. Samuelson (1962: 18) might have been referring to himself when saying that the only reason to engage in economic theory was "our own applause", but nobody lived by that maxim more single-mindedly than Debreu.

As Weintraub (1985: 112) has noted, the most obvious effect of Debreu's interventions into economics has been "a hardening of the hard core" of the Walrasian research programme. Within the framework of a Lakatosian conception of science, the hard core assumes privileged status. It is the part of the theoretical structure that is bracketed off from falsificationist critique, whereby the stipulation of relationships within that part of the theory is taken as given, to reflect the basic propositions that allow the theory to work. Hard core relationships are rendered immune to direct refutation through appeal to data, because their task is not to survive empirical testing but to highlight how empirical tests might be constructed in the "protective belt" of auxiliary hypotheses (Lakatos 1978: 48–51). Debreu's complete disregard for populating his theoretical exposition with anything other than mathematical symbols therefore might not matter in its own terms if he was engaged solely in hard core endeavours that had no implications elsewhere. But they always do, because hard core activities tell the community of practicing researchers what is thinkable for them.

Debreu's writings certainly seem to provide evidence that he isolated hard core matters in his own mind. Arrow accepted Tarski's logical reworking of the philosophical concept of truth to push himself to ever greater heights of mathematical sophistication, through which Cauchian challenges to post-Cauchian proof-making would increasingly disappear (Kemp 2012: 54). Debreu operated to a different standard. He knew that the formal model of the economy he was trying to refine was simply an extended metaphor and that, as such, it could neither be true nor false in any substantive sense (Klamer 1994: 50). It could only be true or false in a formal mathematical sense, bringing back to the discussion Jaakko Kuorikoski's (2021: 201) distinction between applied mathematical explanation and formal mathematical pseudo-explanation. Debreu was adamant that the substantive sense of true and false belonged solely to the descriptive realm and that it was not his job as a mathematical economist to have any views on descriptive reality. The frequent criticism that Debreu's equations bore no resemblance to data-rich understandings of the world around him – Mark Blaug (1994: 131), for instance, accused him of promoting "a fetish of theory" – is therefore meaningless from Debreu's perspective. To do what his critics asked would be to negate his understanding of the very purpose of research. As Christopher Bliss (1993: 227) has noted with great perception: "The near emptiness of general equilibrium theory is a theorem of the theory".

Debreu (1986: 1265, 1266) made a virtue out of the "divorce of form and content" and urged his colleagues to "flaunt the separation". The subsequent restriction of theoretical work to the ever more rigorous elaboration of hard core propositions was the "acid test" by which the rigour of his market models could be judged. The radically underdetermined nature of the general equilibrium approach was thus a sign of its enhanced rigour: the more it was shorn of potential for economic interpretation, the better it became. In the Introduction to his 1959 *Theory of Value*, Debreu (1959: viii) famously insisted: "Allegiance to rigor dictates the axiomatic form of the analysis where the theory, in the strict sense, is logically entirely disconnected from its interpretations". He considered the greatest strength of his approach to be the freedom of expression delivered by the fact that only the axiomatic structure can determine the meaning of the mathematical solution (Baumgärtner, Faber & Schiller 2006: 157). Assumptions can therefore be worked through faultlessly as matters of logic, but there is never a requirement to justify the choice of assumptions (Nadeau 2003: 59). Debreu (1991a: 6) wrote of the need for economic theorists to be "impartial spectators of a play of which they are the actors" (Debreu 1986: 1266). Arrow, remember, had been gently rebuked during the later stages of the drafting of the 1954 *Econometrica* paper for saying that their existence proof should be designed with normative questions in mind (Düppe 2012b: 503).

Unlike Arrow, Debreu had not alighted on the existence problem from a concern with the *economics* of general equilibrium (Debreu 1984: 268). Indeed, his initial exposure to state-of-the-art economic theory while an assistant at the Centre national de la recherche scientifique in the late 1940s had been underwhelming (Weintraub & Mirowski 1994: 260). The available mathematical techniques failed his expectations, and even when his next encounter with economics alerted him to the French tradition of abstractionism – this time through the work of future Nobel laureate Maurice Allais – it was merely the *potential* for mathematical sophistication that he detected (Ingrao & Israel 1990: 281). Rather than viewing Hicks and Samuelson as something to build on economically as Arrow had, he thought that their work was evidence of the obstacles that still needed to be overcome before a precise specification of the solution to the existence problem could be delivered. His journey to the fateful day in January 1952 when the Director of the Cowles Commission, Tjalling Koopmans, gave him a technical report of Arrow's to read, passed instead through the work of the mathematicians von Neumann and Wald. They were responsible, he said, for the only pre-1945 papers in mathematical economics to really deserve the name (Debreu 1986: 1265). Arrow had enjoyed the advantage of personal contact having had Wald as his PhD adviser, but it was Debreu for whom Wald's research really presented the key to the future.

Debreu (1984: 268) gave Wald a special mention in his Nobel Lecture for having reworked Gustav Cassel's own reworking of Walras's system of simultaneous equations. For Debreu, this was the first significant step towards a purely formal axiomatization of general equilibrium. Wald had added a value-theoretic premise to Cassel's equations in a manner directly at odds with Cassel having previously taken one out of Walras's. Samuelson (1938: 61) celebrated Cassel's original excision as it made economics less conceptually reliant on utility considerations, and Arrow seemed content to work within Samuelson's price-theoretical inheritance. By contrast, Debreu (1984: 268) championed Wald's restoration on the grounds that it created clear blue water between a potentially tractable problem in the hard core and a series of intractable problems in the protective belt. The economic content was so far in the background as to be entirely invisible – these were not, after all, hard core propositions as he understood them – because his sole interest was in raising Wald's existence theorem to a new level of mathematical purity (Sent 1999: 730).

Debreu studied mathematics at the École Normale Supérieure in occupied Paris during the Second World War. What really left its mark from that time was the tutorship of Henri Cartan (Baumgärtner, Faber & Schiller 2006: 161). This ensured that he was "trained in the uncompromising rigor of Bourbaki" (Debreu 1984: 268). Nicolas Bourbaki remains a collective of predominantly French mathematicians who, since its founding in 1934, have always worked anonymously behind the pseudonym in attempts to create a purely axiomatic structure for mathematics. It promotes a radically ascetic approach in which everything other than the logical connections of its own internal relationships could be removed from the analysis (Aczel 2007: 127). The Bourbakists first came together in a series of joint publications as a direct response to Gödel's incompleteness theorems (see Chapter 2). Gödel had shown that no complete system of mathematical axioms contains enough information internal to itself to prove its own validity; as a consequence, some undecidable propositions from outside the system have to be allowed in to sustain the impression of validity (Smullyan 1992: 106). Debreu instead followed Bourbaki's line in accepting that mathematical metatheory could concentrate on nothing other than explaining in its own terms the use of abbreviated symbols (Punzo 1991: 5). This was really only to sidestep Gödel's challenge rather than to answer it. However, it allowed the Bourbakists to retreat into a level of mathematical purity that the group's self-appointed spokesperson Jean Dieudonné latterly gave the title, "The Music of Reason", which involved the "progressive aban-donment of the concept of 'evident truths'" (Dieudonné 1992: 203).

The Bourbakists' approach was to start with as few axioms as possible and then to see whether any of these could be removed and still leave the under-lying system of logic intact (Dieudonné 1970: 138). Arrow and Debreu's 1954 *Econometrica* paper follows just such a pattern, whereby the first of the math-ematical proofs of a competitive equilibrium requires only four conditions to hold (Arrow & Debreu 1954: 272). Debreu (1989: 134), moreover, treated subsequent demonstrations that some of the assumptions underpinning these initial conditions could be weakened as a definite advance. Mathematical purity had historically been more highly revered in the French intellectual culture into which Debreu had been socialized, but what started off as a dis-tinctly minority Bourbakist interest in American mathematics departments in the 1940s spread more widely across campus in the 1950s (Martins 2014: 27). Samuel Eilenberg was particularly influential in this trend (Eilenberg & Mac Lane 1945, 1950; Eilenberg & Steenrod 1952). He was known to Debreu and his work was cited as a methodological inspiration in earlier drafts of the famous existence paper. The sense of breaking new ground is evident, though, in Debreu's subsequent insistence that all references to Eilenberg were removed from the published version, because other economists simply would not understand what they signified (Düppe 2012b: 505).

The Bourbaki manifesto was nothing if not an attempt to reclaim territory for mathematics previously ceded to the physical sciences (Aubin 1997: 302). As argued by André Weil (1992: 101), the Bourbakist mathematician who took Debreu under his wing at the Cowles Commission, it was a project designed to allow mathematical developments to arise from within mathematics itself. A radical autonomy was thus claimed by Bourbaki, ensuring that their texts "looked like a storehouse of abstract forms" (Giocoli 2003: 26), a "'Taylor system' for mathematics" (Weintraub 2002: 101). But once conceptualized formally in that way they were free to spread outwards and make a play for subject matter that had previously been studied very differently. Samuelson (1983c: 838), never one to pass up an opportunity to criticize his opponents, complained in the early 1980s about how Bourbakism had "overtaken economic theory".

Their project was one of unification, designed to overcome the fear of disorder contained within the persistence of different mathematical styles (Bourbaki 1950: 221). According to the Bourbaki manual that had most influence upon Debreu's early forays into economics, the "method of reasoning" focuses on "laying down chains of syllogisms" (Bourbaki 1950: 223). Mathematics could thereby become a tool of potentially unlimited applicability: in Dieudonné's (1970: 141) words, "a center from which all the rest unfolds". Debreu's *Theory of Value* was unrelentingly Bourbakist in this regard. For him, the reason for embracing axiomatization was to attempt to rid economic theory of the residual ambiguity of its core concepts (Ingrao & Israel 1990: 287). The joint paper with Arrow had already signalled the need for clarity of mathematical definitions (Arrow & Debreu 1954: 266). But this was nothing compared with what would come later in his sole-authored work. Debreu never gave up on the idea that the ultimate goal was to completely Bourbakize economics, to turn it from a seething hubbub of competing voices to a subject field in which nobody needed to announce their achievements because each advance in theoretical generality would be immediately apparent for what it was (Giocoli 2003: 124).

However, in the process of reconstituting economics as an exercise in creating tools of exposition, the analysis of the *Theory of Value* ceased to focus on anything that genuinely looked like an explanatory model (Clark 1992: 165). The analysis revolved solely around determining the tightest possible structure of language for describing an economy in mathematical terms, where all of the relationships deduced through use of that language must be true by definition. In this way, Debreu did not so much overcome the difficulties entrenched within the Bourbakist objective of mathematical purity as import them directly into economics (Weintraub & Mirowski 1994: 245). As some of the most significant exponents of general equilibrium theory have argued, mathematical proofs "are not results" in any literal sense of the word (von Neumann & Morgenstern 1953: 14), because a purely proof-making economics "is not engaged in description at all" (Hahn 1973a: 323). No mathematical objects constructed to such a formula can ever again be a described function.

Bourbakism, like so many of the other moves in the prehistory of economics imperialism, is founded on the principle of exclusion, and a lot can be learnt about what is at stake when economics is presented as a universal social science template by continually calling these losses to mind. Bourbaki remains a siren figure for those who wish to be seduced by an approach that is "self-speaking, self-contained and invulnerable" (Düppe 2012a: 420), but this always has to be at the cost of the social negotiation of the meaning of practices that shape the experience of social living. However, the search for meaning has not gone away, even amongst those whose methodological Bourbakism suggests they should have known better. Moreover, they have delivered some staggering findings.

THE SHOCKWAVES OF THE SONNENSCHEIN–MANTEL–DEBREU THEOREM

The Arrow–Debreu *Econometrica* paper today occupies a curious double place in the history of economic thought. On matters of mathematical technique, it swept all before it. It was a consciousness defining moment for economics as a whole, challenging the subject field's previous self-image by initiating new standards of precision and rigour that looked like they would quickly become the norm (Weintraub 2002: 101). For the first time, it seemed, it was feasible to state in certain terms that the institutions of the model world could coordinate individual economic activity in a way that economists had thought might be possible for the previous 200 years but had been unable to prove. In that moment, it appeared that the basic mode of enquiry had changed, presumably permanently, and that in future every advance in economic theory would occur under the influence of the axiomatic demands of the formalist revolution. However, as early as the 1960s, it had become obvious that economics was not heading for the future foretold in Arrow and Debreu's existence theorem (Backhouse 1994: 217). Axiomatic proof-making was still the mathematical standard to which the professional community paid homage, the ultimate methodological goal of economic theory, but it was increasingly not what it practiced. Within only ten years of the publication of Arrow and Debreu's revolutionary paper, the professional common sense that had flirted with formal general equilibrium theory had moved back to exemplifying theory, where less exacting demands were placed upon economists' mathematical capabilities (Morgan 2012: 371). A shift was also clearly under way towards more applied research, with the world beyond the model consequently making a comeback in economists' theoretical activities (Backhouse & Cherrier 2017: 5).

Research programmes, though, do have a habit of developing a momentum of their own. Out of sight of most of their colleagues, the mathematical sophisticates continued to explore the outer limits of general equilibrium theory into the 1970s. It was at this time that the most eye-catching revelations were published. Abu Turab Rizvi (2003: 384) has called them "a spectacular series of impossibility results", which on closer inspection knocked down, in cascading fashion, multiple economic claims that it was previously thought were unchallengeable when in possession of a rigorous existence theorem. The desire to focus on the determinacy of the general equilibrium system was forcibly overwritten as an avalanche of results revealed it to be radically indeterminate (Mandler 1999: 41). Amidst the wreckage the mathematical structure continued to stand tall, the original achievement remaining undimmed by the subsequent doubt as to whether it translated into anything at all that was meaningful economically when inferences were drawn from the proof to the theory. The impossibility results show that no behavioural restrictions are sufficient in the world within the model to guarantee that market demand functions operate as the theory requires (Kirman & Koch 1986: 460). Basic market models thus appear to be robbed of their foundational essence (Hahn 1982: 747). Yet if they can no longer be trusted to work in their own terms, then how much faith should be placed in attempts to export their intrinsic mode of reasoning beyond economics and into other social sciences? Are complex mathematical objects that most proponents of economics imperialism would struggle to intrinsically understand therefore threatening to pull the rug from beneath their project?

The mathematical proof of the existence conditions for a competitive equilibrium was a clear advance within a context in which no such proof had previously been attained, but as soon as the existence proof had been established then it was reasonable to ask two further questions (Weintraub 1977: 14). The first was whether the competitive equilibrium exhibited by the world within the model was unique. The presence of multiple equilibria that were mathematically indistinguishable would obviously be unhelpful. It would mean that

market processes were rather more haphazard than the standard theoretical story of allocative efficiency allows them to be. The second question was
whether the competitive equilibrium exhibited by the world within the model
was stable. The possibility that the system might be as likely to be led away as
towards the equilibrium position would obviously also be unhelpful. It would
mean that there were circumstances in which the market mechanism failed
to produce the self-ordering properties that are so frequently ascribed to it.
In a series of independently written papers, Hugo Sonnenschein (1972, 1973),
Rolf Mantel (1974) and Debreu himself (1974) proved conclusively in formal
mathematical terms that the equilibrium exhibited by the world within the
model was neither unique nor stable. Together, these contributions provide
the basis of the Sonnenschein–Mantel–Debreu theorem that demonstrates
the limits of assuming the presence of perfect market-clearing dynamics. The
existence proof, in other words, could stand on its own as an example of what
mathematical purity could bring to economic theory, but it could not be but-
tressed by the further demonstrations that would make it genuinely economically meaningful.

What immediately jumps off the page, of course, is Debreu's presence
amongst those who finally killed off the hope that his own existence theorem might have unlocked the market mysteries which economists had wrestled with throughout the modern age. It is noticeable that the original 1954
Econometrica paper made no attempt to go beyond existence (Arrow & Debreu
1954: 266), and Debreu always dismissed the idea that a simple existence proof
might yield the economically meaningful information that Arrow sought with
his later collaborators (Varian 1984: 7). Debreu entered the debate triggered by
Sonnenschein only to do some mathematical tidying up by showing that there
was greater generality to his findings than had hitherto been demonstrated
(Debreu 1989: 135). In the interim he chose to leave well alone. Wald's work,
which he had always found so inspiring, used exceptionally strong assumptions to suggest what might be there in the equations in addition to existence
(Wald 1951 [1936]: 368–9). Debreu's meticulous attention to the mathematics
in Wald's study seems to have alerted him to the likelihood that the subsidiary
projects to demonstrate uniqueness and stability were going nowhere. Just as
Wald had prediscovered the Arrow–Debreu existence theorem in Debreu's
mind, so in similar fashion it could be said that Debreu prediscovered the
blind alleys down which Wald's intuition had lured economists. At the very

least, his close colleague Werner Hildenbrand (1983: 26) has written that his silence on issues of uniqueness and stability was deliberate from the start.

Sonnenschein (1972: 549) had activated what eventually became the destructive tendency by asking: "Can an arbitrary continuous function, defined on a compact subset C of the interior of a positive orthant, be an excess demand function for some commodity in a general equilibrium economy?" Put more straightforwardly, he was interested in whether the mathematical structure pioneered by Arrow and Debreu failed to rule out patterns of behaviour within the model world that were markedly at odds with the way in which market rationality is assumed to impose itself on properly socialized individuals. As Bruna Ingrao and Giorgio Israel (1990: 315) have noted: "It is hardly necessary to point out that such a question seems framed deliberately to rule out the worst eventuality, i.e., an affirmative answer". What happened next should have become the stuff of legend as multiple and ever more exacting "yeses" ensued. But those who continue to use versions of a market model today, whether as economists or economics imperialists, have found it more beneficial to avert their gaze (Samuels 2007: 171). Sonnenschein's (1973: 353) "yes" showed that an excess aggregate demand function – itself an indication that not all markets are clearing simultaneously even in the model world – could be generated by a Walrasian pure exchange economy if that economy had only a small number of commodities. Mantel (1974: 348) extended the result to a "yes plus" by showing that it also held in the presence of a larger number of commodities. Debreu (1974: 16) then demonstrated the "yes double plus" case of the most general result possible, where there were no restrictions whatsoever on the number of commodities (see also Debreu 1984: 274).

The difficulty revealed in the Sonnenschein–Mantel–Debreu theorem was that pretty much every continuous function on the demand side of the market can be scaled up to an excess demand function. From the perspective of general equilibrium theory this is a "smoking gun" (Hands 2012: 388), a "nightmare … running through all research into uniqueness and stability" (Ingrao & Israel 1990: 317). It opened the Pandora's box of systematic counterexamples to many of the things modern economic theory requires to be true (Hills & Myatt 2010: 72). Since the 1760s, the language of demand and supply has been integral to economists' most basic visualization technique of what a market is and how it can be thought to function (Thweatt 1983: 288). The theory states in the most unequivocal fashion that a single price encapsulates the maximizing solution at which no more nor less would be either demanded or supplied. Debreu's (1974: 21) proof, though, encouraged the idea that there were multiple prices – perhaps even an unlimited number of unique prices – at which demand might be brought into line with supply. As a consequence, the world within the model provided no clue about whether one equilibrium price structure was to be preferred to any other. The most that has subsequently been shown is that these multiple solutions are finite in number. Yet even here there is nothing intrinsically reasonable about the economic restrictions that are required to move away from the worst-of-all-worlds scenario. As Frank Ackerman (2002: 60) has suggested: "Not only does general equilibrium fail to be reliably stable; its dynamics can be as bad as you want them to be".

A face-saving fix has been attempted to preserve the integrity of the most basic market models: allowing one person to stand in for all economic agents (Mirowski 2002: 450). This is a much stronger assumption than any of Jevons, Robbins or Samuelson imposed on the behaviour of market agents, even as they enacted ever greater retreats from realisticness. It is that a single person – the representative individual – populates the market on behalf of everyone in the model, enabling the hypothetical process of market coordination only to have to cope with one set of consumer preferences (DeCanio 2014: 144). Even here, though, Alan Kirman and Klaus-Josef Koch (1986: 458) have shown that the negative Sonnenschein–Mantel–Debreu results can be generalized to a population of nearly identical consumers, nearly identical preferences and nearly identical incomes. There seems to be no way of getting around the fact that the key impossibility results of the 1970s allow an almost unlimited capacity to generate market models whose equilibrium is unworthy of the name (Scarf 1981: 469; Saari 1992: 360). The narrative content of market models can ask the reader to look the other way, but only as a result of deliberate misdirection.

Arrow (1987: 202) has responded to these findings by saying that "you cannot get any information essentially on aggregate demand functions" and Debreu (1986: 1263) that "excessively stringent" assumptions now prevail on anything but existence conditions. Taken together, this amounts to something approaching an admission that the stellar achievement of the original existence theorem actually had no compellingly positive implications. It flattered to deceive because it led to all sorts of contrary examples to the most desired features of general equilibrium. As unrealistic as were the behavioural assumptions that Arrow and Debreu used, they still were not strong enough to allow sufficient structure to be imposed on the model world for it to replicate the essential characteristics of economists' basic theoretical claims (Hands 2012: 380). In other words, it is no longer possible to treat *tâtonnement* as a "harmless 'as if' assumption" in the wake of the probing mathematical studies that have shown it to be a logically inconsistent proposition (Weintraub 1977: 14). As Egbert Dierker (1974: 55) has noted, it is only through introducing a level of structure that turns the system of equations into a purely mathematical game that it becomes possible "to obtain the pretty behavior which economists would like to have" (see also Hahn 1965: 127; Davis 1989: 429). Robert Sugden's (2000: 24) insistence that a model world only has to be thinkable has always looked like something of an anything-goes licence. But post Sonnenschein–Mantel–Debreu, crucial elements of the market model even seem to have slipped those bounds.

Sonnenschein's innocent-sounding question about arbitrary continuous functions eventually showed that almost no reasonable economic claim could be made on the basis of general equilibrium theory (Shafer & Sonnenschein 1982: 675). Those who spent the most time investigating the ensuing "ever cruder reality" (Ingrao & Israel 1990: 328) turned out to be the least likely to make exaggerated economic claims in its name. They displayed critical awareness of how little even the most basic market models can say about market phenomena, let alone the non-market phenomena economics imperialists use it to explain. Axiomatic research thus bred a generation of economists who were fully cognisant of the limits of their profession's commonly agreed theoretical principles. Moreover, the search for a close specification of those limits has been reason enough for them to have continued working in the formalist register, even as their colleagues tended to act as if nothing of any great significance was happening there. As Frank Hahn (1973b: 14–15), one of the doyens of general equilibrium theory, wrote over 50 years ago: "This negative role of the Arrow–Debreu equilibrium I consider almost to be sufficient justification for it, since practical men and illtrained theorists everywhere in the world do not understand what they are claiming … when they claim a beneficent and coherent role for the invisible hand".

CONCLUSION

The word "revolution" is often overused in methodological appraisals of economics. The way in which I set up the concerns of this book could be criticized on these grounds, but that is nothing compared to where the tendency towards a whiggish historical account depicts sudden leaps in the direction of what is considered to be true today. But if the Sonnenschein–Mantel–Debreu theorem were to be incorporated throughout economics then it would seem to have genuinely revolutionary potential. However, its impossibility results are by no means as well known as they deserve to be. Perhaps the highly abstract nature of its mathematics helps to explain why it has not seared itself indelibly on the consciousness of other economists. The training of its main protagonists persuaded them to present the mathematical claims as if they would speak clearly for themselves. Consequently, there is nothing in the surrounding prose where the commentary has noisily drawn attention to the devastating implications contained within the papers. Debreu (1991b: 3), for instance, wrote of how "the austere beauty of mathematics" allowed him to avoid speaking about what his findings implied for the field as a whole. Just think, though, of how unsettling it would be to how mathematical market models might be used if the Sonnenschein–Mantel–Debreu theorem were to overcome its obvious problems with accessibility. There is a good case for saying that the Sonnenschein–Mantel–Debreu theorem is the great unknown in current discussions of the pros and cons of economics imperialism.

Werner Hildenbrand (1983: 19), another of the doyens of general equilibrium theory, suggests that the basic market model's implications are so weak that it is now time for economists to abandon the standard exchange model altogether. What would emerge in its place would obviously be unlike the economics we can find in the subject field's introductory textbooks. Moreover, it could not lead to the economics imperialism with which we are familiar today. Maybe it took a trained mathematician coming to economics as a second discipline to make the point. As Claude Chevalley, one of the original Bourbakists, has argued, the mathematical tradition pioneered in economics by Debreu is like "a very well arranged cemetery with a beautiful array of tombstones" (cited in Guedj 1985: 20). It has always served its major purpose by showing what could not be said rather than what could, hence leaving many historians of economic thought wondering in retrospect whether the Arrow–Debreu project was doomed from the start (Fusfeld 2002: 217). Much might therefore be read into Debreu's ostensible appropriation of Francis Bacon's aphorism, *citius emergit veritas ex errore quam ex confusione*: truth emerges sooner from error than from confusion (Hildenbrand 1983:6). The appeal of Bourbakism to Debreu was that it separated mathematical results from the social process of negotiating their meaning, hence removing all available sources of confusion. Once the veil of inadequate formalization was lifted through the use of axiomatic methods, it was possible to expose the mistake of believing that the standard exchange model might ever be more than an artefact of economists' creative imaginations.

The prehistory of economics imperialism has always been closely tied to the history of the mathematization of the market model. It is important to emphasize, though, that along the way it has also been dragged into some very uncomfortable intellectual territory. Yet none of this ever gets acknowledged by those for whom economics still stands as the ideal to be emulated across the social sciences. The previous four chapters have hopefully shown that two distinct trajectories are discernible within these interlinked histories. On the one hand, the more that economics has attracted trained mathematicians to its ranks, the more precise and the more robust it has become when appraised by mathematical standards. This is what allows economics imperialists to argue that a necessary gain in scientific reputation will follow whenever they make incursions across disciplinary boundaries. On the other hand, the more that trained mathematicians have put their logical skills to work within economics, the less precise and the less robust the basic market model has become when appraised by the empirical standards of the world beyond the model This is what animates the scepticism of the social science subject specialists about what it is exactly that economics imperialists can say about the complex matter of real social phenomena.

If it was not clear before then it should be by now: as it has evolved through the four distinct phases of its prehistory, economics imperialism has become embroiled in an ever more arduous journey with an entirely indistinct destination. What has been gained in matters of technique has almost always been counterbalanced by what has been lost in matters of content. Scientific unification through technique is one thing, but it should not be confused with the demonstration of unification of content. Economics imperialism seems capable only of enacting the former, but its never knowingly undersold PR machine makes persistent claims in relation to the latter. The unification of explanation across market and non-market domains has been greatly facilitated by the enhanced techniques that were increasingly introduced into mathematical economics. However, this would seem to be little more than the apparently indiscriminate overlaying of the market model onto various non-market phenomena. Even then, the most sophisticated mathematical economists have shown that the market model does not work even in its own terms. An obvious question to ask in such circumstances is whether the self-proclaimed successes of the economics imperialists therefore come at too high a cost in terms of forgone social relevance. This is what guides my discussion in the Conclusion.

Conclusion: metamathematical limits to economics imperialism

MULTIPLE MEANINGS OF RIGOUR AND PRECISION

At heart, the debate about economics imperialism might not have moved on very far from when Ralph Souter (1993b: 94) first introduced the notion into social science (see Chapter 1). This was in the 1930s. He argued that rigour and precision looked very different from beyond a mathematical mindset than from within it. They remain prized assets in economics imperialists' rhetorical armoury but in the absence of Souter's reflections on the many meanings they might acquire. Explanation through mathematical analogy within the model will certainly bring additional rigour and precision to understandings of *that* world, but this should not be confused with saying that the world beyond the model is now fully understood. The system of equations will reveal mathematical solutions to what, in essence, are merely mathematical problems. Inductive inference to the world beyond the model involves a leap of faith that even the most aggressive selling of economics imperialism does nothing to overcome. The colonists purport to operate somewhere between the world within substitute models and actual day-to-day experiences, connecting the two in a causal explanation. Yet these are distinct ontological realms that respond to different standards of rigour and precision. Souter recognized this 90 years ago, but the long-forgotten nature of his work shows that his warnings went unheeded.

Souter (1933a: 377–8) had shown that the economists of his day were left with a choice of entering one of two strictly parallel domains: Lionel Robbins's new one or Alfred Marshall's old one. Robbins's is where economics imperialists continue to be positioned today, with rigour and precision being defined in relation to the logically sound specification of the world within the model. Marshall's attracts the critics of economics imperialism, because its definitions of rigour and precision are linked to how well the world within the model captures the characteristics of the empirical realities it is asked to imitate. In modern-day philosophical terms, two different representational relationships between model and target are being invoked: "standing for" the real world in the former, "making present" the real world in the latter, or representing versus re-presenting (Prendergast 2000: 5). Mathematical analogy can replace observational content in Robbins's model worlds and still be epistemically reasonable, but not in Marshall's.

The mathematician who did most to revolutionize economic theory in the 1930s and 1940s was John von Neumann. His work could hardly have been further in theoretical inspiration from Souter's, but he too cautioned against introducing mathematical analogy into previously non-mathematical domains (Szász 2011: 45). In particular, he worried whether it was destined to disconnect the resulting theory from the real phenomena it was supposed to explain (Hargittai & Hargittai 2016: 44). Von Neumann appears to have been the only person present when Kurt Gödel revealed the first of his two incompleteness theorems in 1930 to have immediately understood what he was hearing. The promise of David Hilbert's formalist refounding of the whole of mathematics evaporated in front of his eyes (see Chapter 2). Thereafter, his own mathematical programme, including his transformative forays into mathematical economics, was structured by an empiricist view of logic (Bueno 2016: 226) and a correspondingly sceptical position on the abstract tradition (Ferreirós 1999: 389). Every economist who has attempted to derive a purely axiomatic account of a mathematical market system has cited as a foundational influence von Neumann's research into fixed point theorems, game theory and existence conditions of general equilibrium. Yet in doing so they have overwritten his own methodological concern for "contact with the strivings and problems of the world" (von Neumann 1948, cited in Rédei 2005: 5). To use the metamathematical language that has featured throughout the book, they have subverted his preference for described mathematical functions and have instead inserted in his name defined mathematical functions. But this is what von Neumann (1947: 196) had to say about such a transposition. "As a mathematical discipline travels far from its empirical source, or still more, if it is a second and third generation only indirectly inspired by ideas coming from 'reality', it is beset with very grave dangers. It becomes more and more purely aestheticizing, more and more purely *l'art pour l'art*." The end point of this trajectory, in von Neumann's view, is "much 'abstract' inbreeding".

As the philosopher Jaakko Kuorikoski has argued, it is important to distinguish between mathematics helping in the explanation of empirical phenomena and mathematics *being* the explanation. Mathematical truths provide knowledge of a very different sort to empirical truths. "How can such other-worldly things as mathematical objects, properties, or facts explain something in our contingent, space-timey world?" Kuorikoski (2021: 189) asks. The theorems within the axiomatic structure of the system of equations can only relate to situations imagined into being by the theorist (Clower 1995: 314). The result is that the theorem has become the subject of "de-empirization" (in von Neumann's (1947: 182) terminology) or "analytification" (in Roger Backhouse's (1997: 127)). It is clearly a point against economics imperialism if the imposition of a mathematical market model across disciplinary borders has such effects. Economics imperialists' tendency to dismiss the value of subject specialists' in-depth knowledge of real phenomena, in favour of their own search for supposedly higher-level truths, perhaps tells us all we need to know in this regard. When a mathematical market model is allowed to stand on its own it can provide only what Juha Saatsi (2016: 1065) calls a "thin role" in explanation and Kuorikoski (2021: 208) "formal understanding" that is "easily mistaken for a special kind of explanatoriness". It can still be rigorous and it can still be precise. But this is a particular type of analytical effort that bypasses in-depth empirical knowledge of concrete social situations.

Lying hidden at some depth beneath contemporary controversies regarding economics imperialism, then, is a history of competing images of mathematics. Its invisibility thus far in the debate should not disguise its significance. The derivational unification on which the practice of economics imperialism is founded owes much to the adoption within economic theory of various forms of Lagrange multipliers (see Chapter 1). In the late eighteenth century, Joseph-Louis Lagrange enjoyed considerable success displacing physical analogy from mechanics and replacing it with purely mathematical analogy

(Doyle 2002: 208). Yet by the time economics was taking its first tentative steps towards becoming a mathematical discipline 100 years later, Lagrange's methodological priorities had been largely overwritten by Augustin-Louis Cauchy's very different vision of what mathematics should be (Stedall 2011: 183). Cauchy's dismissal of the idea that it can be known aprioristically that algebraic arguments are true with respect to every conceivable value of their variables acts as a full-frontal assault on Lagrange's assumption that genuine explanation resides in formal manipulation of algebraic series (Grabiner 1984: 111). However, Cauchy's preference for calculability itself came under attack from David Hilbert's formalist programme, which then suffered a setback at the hands of Kurt Gödel's incompleteness theorems. The use of Lagrange multipliers to refashion economists' market models seems to owe much more to the early Hilbert's image of mathematics than to either Cauchy's or Gödel's (see Chapter 2). The prehistory of economics imperialism clearly stretches across a timeframe in which very different images of mathematics took turns to dominate. Is it therefore left looking irredeemably contradictory?

There can be no single historical story that neatly ties together all of the loose ends of the separate phases of the prehistory of economics imperialism. Indeed, each of the primary protagonists featuring in the preceding chapters appears either to have changed their minds on key analytical issues or to have oscillated between incommensurable epistemic views. Stanley Jevons wished to bring concrete numbers to economic theory in a thoroughgoing manner, treating his hedonistic calculus as a real phenomenon that could be calibrated through experimental observation. But he equivocated between whether these numbers should apply to individual economic agents or *l'homme moyen*, Adolphe Quetelet's "average man", and he was therefore left with an underdetermined market model for each person and an overdetermined market model for social aggregates. The later Lionel Robbins attempted to distance his thoughts from those of his younger self by accepting that the decisions individual economic agents make will be influenced by the legal, social, cultural and political institutions of the time and place in which they live. But it was the apriorism to which he subscribed at an earlier stage of his career that left the lasting mark on the economics profession, compared to which his later change of heart hardly registered. Paul Samuelson tried his hardest to dispense with economists' attachment to apriorist reasoning embedded in simplistic utilitarian psychology. But his efforts to reconstitute market models on the basis of mathematical rather than psychological understandings of maximization never really worked, and he was latterly forced into smuggling assumptions about utility back into economic theory. Kenneth Arrow had always thought that the purpose of his investigation of the mathematical principles of general equilibrium had ultimately been misunderstood, and he spent much of his later career identifying institutional factors that frustrated any straightforward application of a market model. Gérard Debreu went one stage further, demonstrating through faultless mathematical logic that it was unreasonable to assume that even the most basic market model could explain anything beyond single decisions in a hypothetical version of the market economy.

Contrast this with the dreams of the most ardent economics imperialists, which usually end with market models explaining everything everywhere across all social domains. The prehistory of economics imperialism stretches back at least a century before the first card-carrying economics imperialists really began to make their case in explicit terms in the 1980s. Yet it is marked by often unfathomable twists and turns, which have stained the canvas on which economics imperialists today paint their visions to such an extent that it is no longer clear what we are being invited to look at.

From the mid-nineteenth century onwards, economists' instincts have typically been oriented towards the adoption of mathematical models. Those who have attempted to hold out have been seen very much to be swimming against the tide. Since the publication of Gödel's incompleteness theorems in the early 1930s, however, it has been far from self-evident how any mathematical model, not just those in economics, might be able to demonstrate its own consistency. Is it through physical corroboration by number, as the model world is systematically redefined each time new observational data reveal its shortcomings? This would seem to place mathematical market models into a metamathematical realm that precedes both Gödel and Hilbert. Is it through analogy to explanatory structures that exert professional authority in other subject fields, whereby real economic phenomena are still expected to have a presence in the market model, even if they are given the mathematical features of other real physical phenomena drawn from outside the economic domain? It is unclear in which metamathematical universe this would position mathematical market models, because presumably that would depend on the precise balance in play between mathematical and physical analogy. Or is it through reduction to crisply defined theorems that reflect in pure mathematical form a series of simple behavioural axioms, through which the explanation is restricted to the world within the model and further inferences to the world beyond must take place on some other basis? This would have to rely on the metamathematics of the pre-Gödel Hilbert and conveniently overlook the style of work more closely associated with the post-Gödel Hilbert. Nothing in the prehistory of economics imperialism points to any simple means of adjudicating between such radically different solutions to the problem of how hypothetical mathematical models represent real economic phenomena, if indeed they are designed to do so at all. A tailored recap of the prehistory detailed in previous chapters demonstrates the point only too well.

EPISTEMOLOGICAL CONTRADICTIONS IN THE PREHISTORY OF ECONOMICS IMPERIALISM

Starting with Jevons, he had a specific mathematical training that brought him to an equally specific answer to how a mathematical model might be able to demonstrate its own consistency and so explain matter beyond itself. He studied at University College London under Augustus De Morgan, whose approach was popular in Britain at that time, but where Britain was very much a mathematical backwater compared to the speed of development taking place elsewhere in Europe (Schabas 1990: 20). De Morgan taught calculus via a focus on infinitesimals, through which the points of most intense mathematical importance were those where the sign changed between two very small numbers in sequence. These were the turning points of differential equations, at which the gradient of the relevant curve is zero. Equalities thus dominated in De Morgan's mathematical models (Panteki 2008: 384). Jevons (1869: 15) attempted to turn the focus on equalities into an entire logical system, of which his account of the hedonistic calculus within economic theory was merely one example (Jevons 2013 [1871/1879]: 303–14). He envisioned the underlying nature of economic decision-making in terms of the balance between the utility derived from consumption and the disutility associated with work (see Chapter 4). The net utility from each individual's various market-based actions at the moment of balance mimicked the turning point of a curve described by De Morgan's approach to differential calculus. However,

Jevons's mathematics looked distinctly old-fashioned even at the time. He had admitted in letters to his family that De Morgan's classes pushed him beyond his capabilities (Black 1972: 130). But elsewhere in Europe mathematicians had already embraced Cauchy's alternative and more complex account of the calculus, building on Jean le Rond d'Alembert, Joseph-Louis Lagrange and Siméon Poisson's pioneering attempts to set it instead within a logic of inequalities (Gratton-Guinness 2002: 693).

Jevons's approach to the calculus might well have positioned him outside mainstream European mathematics but, as Chapter 4 demonstrated, his approach to rigour was very definitely within mainstream European metamathematics, even if less than fully consciously so. Again, Cauchy would seem to be the most significant authority figure, but there is no direct reference to his work in any of Jevons's *Theory of Political Economy*, *Principles of Science* or *Pure Logic*. Nonetheless, Cauchy's insistence that mathematical modelling must always be consistent with physically described concrete results reappears in Jevons's efforts always to ground economic numbers in experimental data. Mathematical rigour within the modelling process was seen almost universally at the start of Jevons's career in terms of the constraints that were placed on the content of model parameters by experimental data (Weintraub 1998: 1841). This was representation as the re-presentation of real-world effects. There is evidence from the early Jevons's (1863: 489) reflections on chemistry, his first intellectual passion, that he had read Poisson's (1842 [1811]) *Treatise of Mechanics*, which paid considerable attention to Lagrange's method of variational calculus (White 2004: 101). But there is nothing in his later economic work to suggest that he understood the mathematical principles underpinning Lagrange's method sufficiently well to use free-standing systems of equations as the foundation for his market models. Jevons's marginalism might well have set the scene for increasingly sophisticated forms of mathematical market models, but he was only interested in described mathematical functions reflecting the conditions of lived economic experiences. Accordingly, the world beyond his models was an obvious presence within his model worlds.

The later Robbins famously questioned the ease with which his younger self could be persuaded to argue against approaches to the market model that were "quasi-literary in manner" (Robbins 1960: 252). Perhaps it was seeing what the likes of Samuelson, Arrow and Debreu had done to the process of hypothetical mathematical modelling in the meantime that elicited the change of heart. The connection between words and the physical reality they can be used to describe continued to matter to Robbins, whereas Samuelson had first cast doubt on the necessity of that connection for economic modelling, and then Arrow and Debreu had rendered it altogether redundant. But this merely means that Robbins's *Essay* must be viewed as sitting uncomfortably between different standards of rigour. It is presented very much as a form of verbal reasoning, in which the argument emerges from how he redefined much of the basis of economic vocabulary, but the automaticity it imposed on economic behaviour hints strongly at the application of an underlying mathematical logic (see Chapter 5). There is no mathematical content in the *Essay*, and yet still the general structure of its argument pattern is unerringly mathematical. An obvious tension is therefore apparent between his theoretical arguments in favour of pure abstraction and, even at the same early stage of his career, his reservation of the greatest praise for the "non-mathematical exposition" of technical issues in economic theory (Robbins 1933: 198).

It is difficult to escape the impression that Robbins was trapped between traditions theoretically as well as epistemologically. His efforts to normalize quintessential but entirely abstract economic behaviour owed much to the belief that Carl Menger had earned a decisive victory over Gustav von Schmoller in the Austro-German *Methodenstreit* of the 1880s. Viewed from such a perspective, the futility of trying to historicize different accounts of economic agency for different times and different places had already been more than adequately demonstrated (Hodgson 2001: 207). However, Austrian economics was largely anti-mathematical in its third-generation stage in Robbins's day, and yet his greatest intellectual respect was reserved for the avowedly mathematical Jevons and Wicksteed. Robbins would not let go of the idea of referential truth claims that he associated with his English predecessors, which contrasted sharply with the apriorism of his Austrian influences (Campagnolo 2010: 312). Jevons and Wicksteed consequently seem to have been forgiven for their mathematical instincts, because behind their systems of equations they were attempting to bring concrete numerical insight to real economic phenomena (Tribe 2015: 83). However, the account of economic agency that emerges from the text of the *Essay* looks to have much more in common with aprioristic Austrian abstraction than Jevons's and Wicksteed's quantity expressed through numbers. Robbins's *Essay* therefore appears to speak in multiple voices at once on the role of mathematization in the development of market models, which hardly brings clarity to the matter at hand.

The search for commonalities of approach gets no easier when turning to Samuelson. His elevation of maximization principles to the status of universal economic explanation is pretty much unthinkable in the absence of Robbins's prior constitution of the proto-axiomatic and very definitely aprioristic economizing subject. Nonetheless, it is impossible to read Samuelson's *Foundations of Economic Analysis* without coming to the conclusion that it is a conscious repudiation of everything that Robbins stood for, both epistemologically and theoretically (see Chapter 6). At the very least, Robbins's preference for economic arguments presented in the form of verbal expression was replaced by the extensive use of systems of equations as reasoning tools (Morgan 2012: 230). Indeed, this substitution was so comprehensive that Samuelson changed not only the look of economic theory but also the very form of the entities it could be called upon to explain. If real economic phenomena did not have the properties to be easily assimilated to a maximization problem solved through the application of variational calculus, then it had to be assumed that they did simply to allow the mathematics to work. Samuelson knew the systems of equations underpinning the explanation of all sorts of physical behaviour better than any economist before him (Mirowski 1991: 223), the product of self-directed learning once Gilbert Bliss's calculus class had opened his still undergraduate eyes to the apparently limitless range of applications that Lagrangian techniques permitted (Backhouse 2017: 165). Here we see the first signs of derivational unification starting unequivocally to crowd out ontological unification in the market models that economists use in theoretical explanation. The mathematical functions running throughout *Foundations* are defined rather than described (Ingrao and Israel 1990: 182).

Despite this, Samuelson (1983 [1947]: 7, 1963: 233, 1983b: 791) continued to make the case that economics needed to be an observational science founded on operationally meaningful theorems. Cauchy's late nineteenth-century views on how mathematical models become rigorous look to be retained in Samuelson's (1955: vi) insistence that the dynamics of any market model must capture the real conduct of real individuals (Carvajalino 2019: 71). But his use of the variational calculus in the d'Alembert–Lagrange–Poisson–Cauchy tradition rendered actual people invisible, concealed within the anonymizing features of the system of equations. Cauchy formalized the definition of the limit of a function in the 1820s, transforming the calculus into the study of the logic of inequalities (Archibald 2008: 123). Samuelson (1983 [1947]: 60, 231, 454) embraced such a shift to develop the internal mathematical features of his market models in terms of Lagrange multipliers. His maximizing models embed the actions of an economizing character type, in line with Robbins's scarcity definition of economics. But a character type is not an actual person, and there is nothing to say that it will reveal the behavioural traits of any known person. The content of his equations allows no concession to making real-world dynamics present within the model, as would be expected of a genuinely observational science. This sees Samuelson's work enter the realm that Uskali Mäki (2001a: 493–8, 2009a: 363–4) describes as "derivational unification".

When we look at the content of the early examples of proto-economics imperialism produced by Anthony Downs (1957), Gary Becker (1957), Duncan Black (1958), James Buchanan and Gordon Tullock (1962), Mancur Olson (1965) and Richard Posner (1972), it is clear that they have much more in common with what appears in Samuelson's *Foundations* than the much more forbidding mathematical structures to be found in the highly formalist work of the general equilibrium theorists Kenneth Arrow, Gérard Debreu, Lionel McKenzie, Leonid Hurwicz and Hirofumi Uzawa (see Arrow & Debreu 1954; McKenzie 1954; Arrow & Hurwicz 1958; Arrow, Block & Hurwicz 1959; McKenzie 1959; Debreu 1959; Uzawa 1960). The same is true today. Market models had already been fully reconstituted in formalist terms so that their mathematical functions could be considered true in and of themselves at least a decade before Stigler (1984) first used the phrase "economics imperialism" as a positive marker of identification. Yet it would be fair to say that these most recent developments in the prehistory of economics imperialism have left no discernible imprint on its subsequent practice. The claims of the imperialists to be legitimate interlopers across social scientific borders rests on the assertion that they are bringing extra clarity of understanding to the world beyond the model. That is, their market models deliver additional explanatory force that is inaccessible to the subject specialist. By contrast, the Hilbertian meta-mathematics underpinning the work of Arrow, Debreu, McKenzie, Hurwicz and Uzawa, as well as Hahn, Sonnenschein, Mantel and Hildenbrand, allows for no such claim.

Samuelsonian mechanical analogy was always in any case suffused with mathematical analogy, but the former was subsequently required to entirely give way to the latter. General equilibrium theory is consistent with the early Hilbert's (1902: 478) formalist assertion that mathematics can be the arbiter of its own validity: that is, it has logically well-ordered structures which enable it to be true in and of itself (Israel 1981: 206). The mathematical embellishment of axiomatic assumptions is a matter solely of the form of the hypothetical mathematical model and is an entirely separate issue to its content. Marshall's "ordinary business of life" definition of economics is therefore *terra incognita* to anyone seeking to bring an Arrow–Debreu world into their model. The wholesale subordination of substantive content to mathematical analogy might therefore be thought to grant market models an autonomy they otherwise would not have, but this is also a separation from the real economic content we might *want* it to have. A market model constructed in line with Hilbertian metamathematics re-presents nothing of the world beyond; moreover, it is not intended to. It reflects merely what is considered to be appropriate economic theory.

Arrow and Debreu's (1954) existence proof for the conditions of general equilibrium in a hypothetical market system is still a remarkable intellectual achievement. However, it is necessary to be clear what it is an achievement *of*, on which point it seems sensible to defer to the authors themselves. Arrow (1987: 198) says that being able to operate at such a sophisticated level of mathematical enquiry forces economists to be more nuanced in their specification of economic terms. Whatever else might subsequently have been claimed of it, Arrow suggests that the most potent use of the existence proof is to call the bluff of those who make exaggerated claims on behalf of economic theory behind the veneer of only superficially plausible mathematical models. This is the use of high-powered mathematics to prevent economists succumbing to ungrounded mathematical performance. Debreu (1991a: 4) pursues a very similar argument. He suggests that the existence proof liberates the underlying conceptions of consumer and producer behaviour from their systematic misspecification in standard market models of demand and supply. Werner Hildenbrand (1983: 19) cuts to the chase in arguing that Arrow, Debreu and their like have shown the implications of the basic market model to be so weak in their economic meaning that it would have been the right choice to have abandoned it altogether following the rush of impossibility theorems published in the 1970s (see Chapter 7).

FINAL WORDS

Here we reach, at last, the crux of the epistemological problems associated with the contemporary practice of economics imperialism. Stigler kickstarted the process of economics imperialists' bold declarations that their mathematical market models can explain everything everywhere *only after* Hildenbrand had said the mathematical evidence confirms that they can explain next to nothing even within economic theory itself. Intemperate warnings to other social scientists to prepare for market models to sweep all before them post-dated a more mildly mannered warning from a renowned authority figure on mathematical economics that economists should be going back to the drawing board to rebuild their market models from scratch.

The difficulty for economics imperialism is this. It relies on the idea – drawn implicitly from Hilbertian metamathematics – that mathematical market models based on pure exchange relations contain mathematical truths that translate into broader social truths, so that those market models in turn exhibit universal explanatory potential. Those, though, who have explicitly adopted the Hilbert programme within economics have argued that the only mathematical truths to emerge from it should have been enough to extinguish the last vestiges of faith in even the most elemental market model. Economics imperialism also relies on the idea – this time drawn from Cauchian metamathematics – that mathematical market models carry within them the content of lived experiences from various domains as they are introduced across the social sciences. However, neither do they nor can they. Mathematical market models merely reflect back generally agreed economic principles into theoretical accounts of how the world would work were it to provide a perfect template for the theory. Cauchian metamathematicians are hesitant when an obvious gap opens up between the world within the model and the world beyond, when the features of real phenomena are absent from the way they are represented mathematically. But the message from mathematical economists working in this tradition is just how hard it is to introduce genuinely lived experiences into even the most basic version of the market model. By contrast, Hilbertian metamathematicians are comfortable preserving such a gap so that mathematical functions can be presented with logical precision. But the message from mathematical economists working in this tradition is not to use the standard market model at all. All told, this leaves economics imperialists looking like they inhabit an epistemological no-man's-land.

Economics imperialism might therefore reduce simply to a state of mind. A wide variety of well-populated literatures might help to explain its origins: the professionalization of economics; the sociology of expertise; the institutionalization of ideas; the effects of paradigmatic thinking. All point to the likelihood that it is easier to defend certain ways of doing economics than others, whereby someone who begins with a mathematical market model is almost certainly not going to be pressed to justify why they start from there, but anyone else might well have to explain each of their founding assumptions one by one. It is a small step from recognizing how certain modes of explanation are professionally privileged to the feeling that market models reveal some ultimate truth underpinning the organization of human affairs. If your trained intuition operates in this manner, the related idea that you can now explain any and, indeed, every legal, social, cultural or political phenomenon must be entirely exhilarating. There is a thrill-seeking element to economics imperialism that manifests in the delight that its proponents take in their declarations that they have arrived in other social scientific subject fields both unannounced and unsolicited. The flipside of this, of course, is seeing subject specialists take fright at the thought that their expert knowledge will consequently suffer wholesale devaluation.

However, it is noticeable how work on the basic market model has changed since the 1970s. The formalists' impossibility theorems may have left no discernible mark on economics imperialism, but the same is not true of the underlying models on which its practices are based. There was no revolutionary moment in which all desire to work with market models was jettisoned (see Chapter 7), but the *nature* of those models is now considerably more contested. The investigation of contentless market models dominated the intellectual context in which the earliest pre-emptions of explicit economics imperialism took place in the 1950s, 1960s and 1970s. Yet these models now by no means stand alone as background influences, as there are many more forms of economic theory from which to choose (Fine & Milonakis 2009: 57). The first three stages of the prehistory of economics imperialism were animated by how far economists were prepared to go to banish from the dominant account of the market model anything other than that which they could say was a fundamental economic essence (see Chapters 4–6). But the stipulation

of suitably historicized and suitably contextualized socio-psychological predicates of economic behaviour is now once again back in fashion (Frey & Benz
2004: 62). Content has returned as economic theory has experienced its own
inhaltliche moment (see Chapter 2). Behavioural economics, experimental
economics, complexity economics, feminist economics, ecological economics, evolutionary economics: all these approaches and more besides are seeking to dethrone the simple application of formal mathematical models. The
applied turn that has been ongoing since the time of the formalist impossibility theorems has also been gathering pace recently (Backhouse & Cherrier
2017: 19). Space constraints do not allow for a full review, but all of these
are content-rich alternatives to the uneconomic economics of mathematical
market models, where the content is an attempt to make the real world present
again within the explanatory model.

The fracturing of the market model's hegemony *within* economics certainly
makes it seem that the project of economics imperialism has been mis-sold.
Even if it was never verbalized explicitly like this, the proponents of colonization were saying that boundary-hopping market models met the criterion
of ontological unification. That is, they were able to explain the cause of all
action in the same way, because the cause was always the same. However, the
Sonnenschein–Mantel–Debreu theorem showed that demand functions are
entirely unconstrained in their shape (Petri 2021: 313), and many economists
have responded by asking how real people enact different forms of behaviour
in different markets (Reis 2018: 134). The aim is to bring the complexities
of in-time real-world decision-making back into the model and therefore to
accept that there should be ontological constraints on explanatory unification
(Kuorikoski & Lehtinen 2010: 347).

By contrast, the most economics imperialists can aim for is integrating the
way in which sense is made of the world. This is obviously an explanation of
sorts, but not in terms of revealing actual causal dynamics of actual human
processes (Davis 2016: 88). It is about finding a template for understanding that
passes elementary logical tests but also bypasses the need for the explanation to
be based on knowledge of real economic phenomena (Hodgson 1994: 22; Fine
2000: 15). The introduction into economists' market models of the Lagrangian
mindset should have been a clue even at the time that ontological constraints
on explanatory unification were being downgraded. Lagrangian formalism
makes it possible for the mathematical logic of the system of equations to

overwhelm in the explanation the physical phenomena they purport merely to represent (Blanchard & Brüning 1992: 6). Market models henceforth have responded in the first instance to the mathematics through which they are visualized theoretically. The world being made sense of here is most obviously that of economic theory rather than of everyday experiences. Even then, there are now more versions of economic theory struggling for dominance than there were in the post-Second World War heyday of the Lagrangian mindset. At least as many of these want to rehabilitate the ontological constraint on explanatory unification as remain committed to the *status quo ante*.

Economics imperialists therefore seem to be satisfying themselves with mere derivational unification. Theirs is unification as a formal possibility, rather than unification as factual discovery (Mäki 2001a: 503). This does not mean it has no role within the social sciences, only that care must be taken not to overstate what that role might be. Unification as a formal possibility is based on common mathematical specification of a variety of legal, social, cultural and political problems once all the starting propositions have been stipulated (Witt 2009: 366). It is not based on the generation of deep subject specific specialist knowledge of the conditions under which those problems come to life. Indeed, such knowledge is almost certainly an impediment to the onwards march of economics imperialism, because it can be relied upon to reveal the complex difference of the relevant real phenomena, whereas economics imperialism only offers a single template for understanding. This is presumably why the loudest advocates of the practice of economics imperialism appear so brazen in their dismissal of the value of such expert knowledge (Hirshleifer 1985: 61–2; Lazear 2000: 100, 140). Yet this may be no more than an acknowledgement that market models are the equivalent of Hilbert's thought-things when applied to non-market social phenomena: they can be reasoned with but should not be mistaken for something real.

Is this, then, all that the achievements of the revolutionaries amount to as they push back the frontiers of their advances across the social sciences? Is their brave new world only that of formal understanding that casts more light on the features of the model used in the explanatory activity than it does on what the model is being used to explain? In this case, it would qualify only as what Mäki (2009a: 373) calls "bad" economics imperialism (see Chapter 3). In his view, the only epistemological basis on which it might be possible to justify individual cases of economics imperialism is when ontological unification is secured. Every other instance is epistemologically illegitimate, whatever epistemic insights might be gained from assessments of how hypothetical mathematical models fare in relation to their self-made worlds (Mäki 2009a: 366). True explanatory progress, rather than the pseudo-explanations of formal mathematical reasoning, does not follow from mere derivational unification.

This knowledge on its own, though, is unlikely to offer sufficient reassurance to any subject specialist whose area of expertise has already been redefined by colonizers brandishing their market models. It is also unlikely to stop others from fearing that their home within the literature is the next to be invaded. We thus come to the final irony. There is nothing in any of the previous chapters that serves as an obvious justification for the claims of the imperialists, with the real contribution of market-model thought-things only being revealed in the presence of rather more circumspect claims about explanation. Both the philosophical and the historical arguments line up behind those who wish to return their specialist literatures to what they were prior to the onset of colonizing activities. At the very least, the work of the economics imperialists should not be seen as a direct replacement for their own, because it exists on a fundamentally different ontological plane. Yet it is the people who have the arguments on their side who also find themselves repeatedly on the defensive. We therefore should not expect the economics imperialism trend to simply fizzle out, and fears of its acceleration remain very real. There is always a tendency for debates about how the respective explanations work to dissolve into disputes about which is superior. This makes it likely that many more appraisals of economics imperialism will be required in the future.

Timeline of developments in economic theory, economics imperialism and the philosophy and mathematics of explanation

Economic Theory	Economics Imperialism	Philosophy and Mathematics of Explanation
		1687: Newton's *Principia* unifies Kepler/Galileo
		1788: Lagrange's *Mécanique Analytique*
		1821: Cauchy's *Cours d'Analyse*
		1842: Quetelet's *Treatise on Man*
1854: Gossen's *Entwickelung der Gesetze*		
1855: Jennings's *Natural Elements*		
		1861–62: Maxwell's "On Physical Lines of Force"
1862: Cliffe Leslie's "The Love of Money"		
1862: Jevons's unsuccessful BAAS paper		
		1865: Maxwell's "Dynamical Theory"
1871. Jevons's *Theory of Political Economy*		
1871: Menger's *Grundsätze*		
		1873: Schröder's treatise on ordinal numbers
1874: Walras's *Éléments*		

Economic Theory	Economics Imperialism	Philosophy and Mathematics of Explanation
		1883: Cantor formalises ordinal number theory
1883–84: Austro-German *Methodenstreit*		
		1887: Hertz proves Maxwell's theory
1890: Marshall's *Principles of Economics*		
1890: John Neville Keynes's *Scope and Method*		
		1891: Hilbert's railway station epiphany
		1892: Lyapunov's thesis on stability of motion
		1899: Hilbert's *Grundlagen der Geometrie*
1900: Pareto calls economics "science of choice"		
		1902: Gibbs's finite-difference probability
1906: ordinal number scale enters economics		
1910: Wicksteed's *Common Sense*		
		1927: Haberler's index number technique
		1930: Gödel's incompleteness theorems
1932: Robbins's *Essay*		
1933: Samuelson took Bliss's calculus class		
	1933: Souter's first use of "economic imperialism"	
1934: Hicks and Allen's indifference curves		
		1934: Founding of Nicolas Bourbaki collective
		1934: Hilbert's *Grundlagen der Mathematik*
1935: Tinbergen first use of the word "model"		

Economic Theory	Economics Imperialism	Philosophy and Mathematics of Explanation
1936: Wald's proto-existence proofs		
		1939: Konüs's application of index numbers
		1941: Kakutani's fixed point theorem
		1941: Tarski's "On the Calculus of Relations"
1947: Samuelson's *Foundations*		
		1948: von Neumann's scepticism of abstraction
		1949: Kneale's theory of explanatory unification
		1950: Slater's "Lagrange Multipliers Revisited"
1954: McKenzie's general equilibrium paper		
1954: Arrow and Debreu's existence proof		
	1957: Downs's *Economic Theory of Democracy*	
	1957: Becker's *Economics of Discrimination*	
	1958: Black's *Theory of Committees and Elections*	
	1962: Buchanan and Tullock's *Calculus of Consent*	
	1965: Olson's *Logic of Collective Action*	
		1966: Hesse's theory of analogical reasoning
	1968: Boulding's AEA President's address	
1969: Tinbergen's Nobel Prize		
1970: Samuelson's Nobel Prize		
	1972: Posner's *Economic Analysis of Law*	
1972: Arrow's Nobel Prize		
1972: Hicks's Nobel Prize		

Economic Theory	Economics Imperialism	Philosophy and Mathematics of Explanation
1972–74: Sonnenschein–Mantel–Debreu theorem		
		1973: Lewis on non-unification of nature
		1976: Popper on situational analysis in economics
	1978: Coase's warning re economics imperialism	
		1980: Nowak's method of idealisation
		1981: Kitcher's general argument patterns
	1982: Stigler's Nobel Prize	
		1983: Leamer's methodology of robustness
1983: Debreu's Nobel Prize		
	1983: Hildenbrand says market model obsolete	
	1984: Stigler's celebratory *SJE* article	
	1985: Hirshleifer's celebratory *AER* article	
	1988: Becker's interview with Swedberg	
	1992: Becker's Nobel Prize	
1995: Lucas's Nobel Prize		
		2000: Morrison's *Unifying Scientific Theories*
		2000: Sugden's credible worlds theory
	2000: Lazear's celebratory *QJE* article	
		2001: Mäki's ontological/derivational unification
	2009: Mäki's good vs bad economics imperialism	

References

Ackerman, F. (2002). "Flaws in the foundation: consumer behavior and general equilbrium theory". In E. Fullbrook (ed.), *Intersubjectivity in Economics: Agents and Structures*, 56–70. London: Routledge.

Aczel, A. (2007). *The Artist and the Mathematician: The Story of Nicolas Bourbaki, the Genius Mathematician Who Never Existed*. New York: Basic Books.

Agassi, J. (2008). *Science and its History: A Reassessment of the Historiography of Science*. New York: Springer.

Akerlof, G. (1976). "The economics of caste and of the rat race and other woeful tales". *Quarterly Journal of Economics* 90(4): 599–617.

Alexandrova, A. (2006). "Connecting economic models to the real world: game theory and the FCC spectrum auctions". *Philosophy of the Social Sciences* 36(2): 173–92.

Aliprantis, C. (1996). *Problems in Equilibrium Theory*. New York: Springer.

Allen, R. (1934a). "The nature of indifference curves". *Review of Economic Studies* 1(2): 110–21.

Allen, R. (1934b). "A reconsideration of the theory of value: Part II. A mathematical theory of individual demand functions". *Economica* 1(2): 196–219.

Amariglio, J. & D. Ruccio (2001). "From unity to dispersion: the body in modern economic discourse". In S. Cullenberg, J. Amariglio & D. Ruccio (eds), *Postmodernism, Economics and Knowledge*, 143–65. London: Routledge.

Ambirajan, S. (1995). "The delayed emergence of econometrics as a separate discipline". In I. Rima (ed.), *Measurement, Quantification and Economic Analysis: Numeracy in Economics*, 198–211. London: Routledge.

Andvig, J. (1991). "Verbalism and definitions in interwar theoretical macroeconomics". *History of Political Economy* 23(3): 431–55.

Archibald, T. (2008). "The development of rigor in mathematical analysis". In T. Gowers, J. Barrow-Green & I. Leader (eds), *The Princeton Companion to Mathematics*, 117–29. Princeton, NJ: Princeton University Press.

Ardley, G. (1968). *Berkeley's Renovation of Philosophy*. The Hague: Martinus Nijhoff.

Arianrhod, R. (2005). *Einstein's Heroes: Imagining the World through the Language of Mathematics*. Oxford: Oxford University Press.

Arrow, K. (1972). "Models of job discrimination". In A. Pascal (ed.), *Racial Discrimination in Economic Life*, 83–102. Lexington, MA: D. C. Heath.

Arrow, K. (1983). *The Collected Papers of Kenneth J. Arrow, Volume 1 – Social Choice and Justice*. Cambridge, MA: Harvard University Press.

Arrow, K. (1987). "Oral history I: an interview". In G. Feiwel (ed.), *Arrow and the Ascent of Modern Economic Theory*, 191–242. London: Macmillan.

Arrow, K. (2002). "The genesis of 'optimal inventory policy'". *Operations Research* 50(1): 1–2.

Arrow, K. & G. Debreu (1954). "Existence of an equilibrium for a competitive economy". *Econometrica* 22(3): 265–90.

Arrow, K. & F. Hahn (1971). *General Competitive Analysis*. San Francisco, CA: Holden Day.

Arrow, K. & L. Hurwicz (1958). "On the stability of the competitive equilibrium, I". *Econometrica* 26(4): 522–52.

Arrow, K., H. D. Block & L. Hurwicz (1959). "On the stability of the competitive equilibrium, II". *Econometrica* 27(1): 82–109.

Aubin, D. (1997). "The withering immortality of Nicolas Bourbaki: a cultural connector at the confluence of mathematics, structuralism, and the Oulipo in France". *Science in Context* 10(2): 297–342.

Auxier, R. & G. Herstein (2017). *The Quantum of Explanation: Whitehead's Radical Empiricism*. Abingdon: Routledge.

Aydinonat, E. (2015). "The two images of economics: why the fun disappears when difficult questions are at stake?". In E. Aydinonat & J. Vromer (eds), *Economics Made Fun: Philosophy of the Pop-Economics*, 55–70. Abingdon: Routledge.

Backhaus, J. & H. Maks (2006). "From Walras to Pareto: introduction". In J. Backhaus & H. Maks (eds), *From Walras to Pareto*, 1–8. New York: Springer.

Backhouse, R. (1994). *Economists and the Economy: The Evolution of Economic Ideas*. Second edn. Oxford: Blackwell.

Backhouse, R. (1997). *Truth and Progress in Economic Knowledge*. Cheltenham: Edward Elgar.

Backhouse, R. (1998a). "If mathematics is informal, then perhaps we should accept that economics must be informal too". *Economic Journal* 108(451): 1848–58.

Backhouse, R. (1998b). "The transformation of US economics, 1920–1960, viewed through a survey of journal arrticles". *History of Political Economy* 30(Supp.): 85–107.

Backhouse, R. (2002). *The Penguin History of Economics*. London: Penguin.

Backhouse, R. (2009). "Robbins and welfare economics: a reappraisal". *Journal of the History of Economic Thought* 31(4): 474–84.

Backhouse, R. (2012). "The rise and fall of Popper and Lakatos in economics". In U. Mäki (ed.), *Philosophy of Economics*, 25–48. London: Elsevier.

Backhouse, R. (2017). *Founder of Modern Economics: Paul A. Samuelson – Volume 1: Becoming Samuelson, 1915–1948*. Oxford: Oxford University Press.

Backhouse, R. & B. Cherrier (2017). "The age of the applied economist: the transformation of economics since the 1970s". *History of Political Economy* 49(Supp.): 1–33.

Backhouse, R. & S. Durlauf (2009). "Robbins on economic generalizations and reality in the light of modern econometrics". *Economica* 76(Supp.): 873–90.

Backhouse, R. & S. Medema (2009a). "Robbins's *Essay* and the axiomatization of economics". *Journal of the History of Economic Thought* 31(4): 485–99.

Backhouse, R. & S. Medema (2009b). "Defining economics: the long road to acceptance of the Robbins definition". *Economica* 76(Supp.): 805–20.

Bailer-Jones, D. (2003). "When scientific models represent". *International Studies in the Philosophy of Science* 17(1): 59–74.

Balisciano, M. & S. Medema (1999). "Positive science, normative man: Lionel Robbins and the political economy of art". *History of Political Economy* 31(Supp.): 256–84.

Barruchello, G. (2012). "Pareto's rhetoric". In J. Femia & A. Marshall (eds), *Vilfredo Pareto: Beyond Disciplinary Boundaries*, 153–76. Aldershot: Ashgate.

Basile, A. & M. Li Calzi (2004). "Who said that a mathematician cannot win the Nobel Prize?". In M. Emmer (ed.), *Mathematics and Culture I*, 109–20. Trans. E. Moreale. London: Springer.

Baumgärtner, S., M. Faber & J. Schiller (2006). *Joint Production and Responsibility in Ecological Economics: On the Foundations of Environmental Policy*. Cheltenham: Edward Elgar.

Baumol, W. (1984). "Foreword". In L. Robbins (1984 [1935]), *An Essay on the Nature and Significance of Economic Science*, vii–ix. Third edn. London: Macmillan.

Becchio, G. (2009). "The genesis of the half-published Viennese autobiography of Karl Menger (1923–1938): new light on the Vienna Circle and the mathematical colloquium". In G. Becchio (ed.), *Unexplored Dimensions: Karl Menger on Economics and Philosophy (1928–1938)*, 1–20. Bingley: Emerald.

Becker, G. (1957). *The Economics of Discrimination*. Chicago, IL: University of Chicago Press.

Becker, G. (1968). "Crime and punishment: an economic approach". *Journal of Political Economy* 76(2): 169–217.

Becker, G. (1976). *The Economic Approach to Human Behavior*. Chicago, IL: University of Chicago Press.

Becker, G. (2007 [1971]). *Economic Theory*. With the assistance of M. Grossman and R. Michael. New Brunswick, NJ: Transaction.

Becker, G. (2010). "Economic imperialism. Interview: Nobel Laureate Gary Becker". *Religion and Liberty* 3(2). https://www.acton.org/pub/religion-liberty/volume-3-number-2/economic-imperialism.

Beed, C. & O. Kane (1992). "What is the critique of the mathematization of economics?". *Kyklos* 44(4): 581–612.

Beinhocker, E. (2006). *The Origin of Wealth: Evolution, Complexity, and the Radical Remaking of Economics*. London: Random House.

Beller, M. (2001). "Born's probabilistic interpretation: a case study of 'concepts in flux'". In P. Galison, M. Gordin & D. Kaiser (eds), *The History of Modern Physical Science in the Twentieth Century – Volume 4, Quantum Histories*, 227–50. London: Routledge.

Berkeley, G. (1754 [1734]). *The Analyst; Or, A Discourse Addressed to an Infidel Mathematician*. Second edn. London: Tonson & Draper.

Berkeley, G. (1871 [1707–08]). "Commonplace Book of Occasional Metaphysical Thoughts". In A. Fraser (1871), *The Works of George Berkeley, D.D., including many of his writings hitherto unpublished: Volume IV*, 419–502. Oxford: Clarendon Press.

Berkeley, G. (1992 [1721]). *De Motu and the Analyst*. Edited and translated by D. Jesseph. Dordrecht: Kluwer.

Bernoulli, D. (1954 [1738]). "Exposition of a new theory on the measurement of risk". Trans. L. Sommer. *Econometrica* 22(1): 23–36.

Black, D. (1958). *The Theory of Committees and Elections*. Cambridge: Cambridge University Press.

Black, J., N. Hashimzade & G. Myles (2012). *A Dictionary of Economics*. Fourth edn. Oxford: Oxford University Press.

Black, R. (1970). "Introduction". In William Stanley Jevons (1970 [1871]), *The Theory of Political Economy*, 7–38. Harmondsworth: Pelican.

Black, R. (1972). "Jevons, Bentham and De Morgan". *Economica* 39(154): 119–34.

Black, R. (1973a). *Papers and Correspondence of William Stanley Jevons, Volume II: Correspondence, 1850–1862*. London: Macmillan.

Black, R. (1973b). *Papers and Correspondence of William Stanley Jevons, Volume III: Correspondence, 1863–1872*. London: Macmillan.

Black, R. (1973c). *Papers and Correspondence of William Stanley Jevons, Volume V: Correspondence, 1879–1882*. London: Macmillan.

Black, R. (1973d). "W. S. Jevons and the foundation of modern economics". In R. Black, A. Coats & C. Goodwin (eds), *The Marginal Revolution in Economics: Interpretation and Evaluation*, 98–112. Durham, NC: Duke University Press.

Black, R. (1982). "The papers and correspondence of William Stanley Jevons: a supplementary note". *The Manchester School* 50(4): 417–28.

Blakely, J. (2020). "How economics becomes ideology: the uses and abuses of rational choice theory". In P. Róna & L. Zsolnai (eds), *Agency and Causal Explanation in Economics*, 37–52. London: Springer.

Blanchard, P. & E. Brüning (1992). *Variational Methods in Mathematical Physics: A Unified Approach*. Trans. G. Hayes. London: Springer.

Blatt, J. (1983). "How economists misuse mathematics". In A. Eichner (ed.), *Why Economics is Not Yet a Science*, 166–86. New York: M. E. Sharpe.

Blaug, M. (1990). "Comment on O'Brien's 'Lionel Robbins and the Austrian connection'". In B. Caldwell (ed.), *Carl Menger and His Legacy in Economics*, 185–8. Durham, NC: Duke University Press.

Blaug, M. (1992). *The Methodology of Economics: Or How Economists Explain*. Second edn. Cambridge: Cambridge University Press.

Blaug, M. (1994). "Why I am not a constructivist: confessions of an unrepentant Popperian". In R. Backhouse (ed.), *New Directions in Economic Methodology*, 109–36. London: Routledge.

Blaug, M. (1997). *Economic Theory in Retrospect*. Fifth edn. Cambridge: Cambridge University Press.

Blaug, M. (2009). "A symposium on *The Nature and Significance of Economic Science* by Lionel Robbins: foreword". *Journal of the History of Economic Thought* 31(4): 417–20.

Bliss, C. (1993). "Oil trade and general equilibrium: a review article". *Journal of International and Comparative Economics* 2(2): 227–42.

Block, W. (1993). "The economist as detective: reflections on Gary Becker's Nobel prize". *Mises Institute: Mises Daily Articles*, 16.02.2010, https://mises.org/library/economist-detective-relections-gary-beckers-nobel-prize.

Blumenthal, O. (1935). "Lebensgeschichte". In D. Hilbert (1970), *Gesammelte Abhandlungen, Volumen III*, 388–429. Berlin: Springer.

Bockman, J. (2011). *Markets in the Name of Socialism: The Left-Wing Origins of Neoliberalism*. Stanford, CA: Stanford University Press.

Boland, L. (2003). *Foundations of Economic Method: A Popperian Perspective*. Second edn. London: Routledge.

Borglin, A. (2004). *Economic Dynamics and General Equilibrium: Time and Uncertainty*. London: Springer.

Bos, H. (1980). "Newton, Leibniz and the Leibnizian tradition". In I. Grattan-Guinness (ed.), *From the Calculus to Set Theory, 1630–1910: An Introductory History*, 49–93. Princeton, NJ: Princeton University Press.

Boulding, K. (1969). "Economics as a moral science". *American Economic Review* 59(1): 1–12.

Boumans, M. (1997). "Lucas and artificial worlds". *History of Political Economy* 29(Supp.): 63–90.

Boumans, M. (1999). "Built-in justification". In M. Morgan & M. Morrison (eds), *Models as Mediators: Perspectives on Nature and Social Science*, 66–96. Cambridge: Cambridge University Press.

Boumans, M. (2005). *How Economists Model the World into Numbers*. London: Routledge.

Bourbaki, N. (1950). "The architecture of mathematics". *American Mathematical Monthly* 57(2): 221–32.

Boylan, T. & P. O'Gorman (2008). "Popper, economic methodology and contemporary philosophy of science". In T. Boylan & P. O'Gorman (eds), *Popper and Economic Methodology: Contemporary Challenges*, 5–32. London: Routledge.

Breger, H. (2000). "Tacit knowledge and mathematical progress". In E. Grosholz & H. Breger (eds), *The Growth of Mathematical Knowledge*, 221–30. Dordrecht: Kluwer.

Breidert, W. (2005). "Berkeley's defence of the infinite God in contrast to the infinite in mathematics". In T. Koetsier & L. Bergmans (eds), *Mathematics and the Divine: A Historical Study*, 499–508. Amsterdam: Elsevier.

Breit, W. & B. Hirsch (2004). *Lives of the Laureates: Eighteen Nobel Economists*. Fourth edn. Cambridge, MA: MIT Press.

Breit, W. & B. Hirsch (2009). *Lives of the Laureates: Twenty-Three Nobel Economists*. Fifth edn. Cambridge, MA: MIT Press.

Bressoud, D. (2008). *A Radical Approach to Lesbegue's Theory of Integration*. Cambridge: Cambridge University Press.

Brock, W. & D. Colander (2005). "Complexity, pedagogy, and the economics of muddling through". In M. Salzano & A. Kirman (eds), *Economics: Complex Windows*, 25–42. London: Springer.

Brockway, G. (2001). *The End of Economic Man: An Introduction to Humanistic Economics*. Fourth edn. New York: Norton.

Brown, A. & D. Spencer (2012). "The nature of economics and the failings of the mainstream: lessons from Lionel Robbins's *Essay*". *Cambridge Journal of Economics* 36(4): 781–98.

Bruni, L. & F. Guala (2001). "Vilfredo Pareto and the epistemological foundations of choice theory". *History of Political Economy* 33(1): 21–49.

Bruno, S. (2010). "Optimisation and 'thoughtful conjecturing' as principles of analytical guidance in social decision making". In M. Faggini & C. Vinci (eds), *Decision Theory and Choices: A Complexity Approach*, 37–63. New York: Springer.

Buchanan, J. & G. Tullock (1962). *The Calculus of Consent: Logical Foundations of Constitutional Democracy*. Ann Arbor, MI: University of Michigan Press.

Bueno, O. (2016). "Von Neumann, empiricism and the foundations of quantum mechanics". In D. Aerts *et al.* (eds), *Probing the Meaning of Quantum Mechanics: Superpositions, Dynamics, Semantics and Identity*, 192–230. Singapore: World Scientific.

Butler, H. & J. Klick (eds) (2018). *History of Law and Economics*. Cheltenham: Edward Elgar.

Cairnes, J. (2004 [1857/1875]). *The Character and Logical Method of Political Economy*. In T. Boylan & T. Foley (eds), *John Elliot Cairnes: Collected Works, Volume I*. London: Routledge.

Caldwell, B. (1982). *Beyond Positivism: Economic Methodology in the Twentieth Century*. London: Unwin Hyman.

Campagnolo, G. (2010). *Criticisms of Classical Political Economy: Menger, Austrian Economics and the German Historical School*. London: Routledge.

Cannan, E. (1922). *Wealth: A Brief Explanation of the Causes of Economic Welfare*. London: P. S. King.

Cannan, E. (1932). "*An Essay on the Significance of Economic Science*, by Lionel Robbins". *Economic Journal* 42(167): 424–7.

Cantor, G. (1883). *Grundlagen einer allgemeinen Mannigfaltigkeitslehre*. Leipzig: Teubner.

Carpenter, W. (1852). "On the relation of mind and matter". *British and Foreign Medio-Chirurgical Review* 10(Oct): 506–18.

Carpenter, W. (2013 [1855]). *Principles of Human Physiology*. Fifth edn. Charleston, SC: Nabu Press.

Carrillo, N. & T. Knuuttila (2021). "An artifactual perspective on idealization: constant capacitance and the Hodgkin and Huxley model". In A. Cassini & J. Redmond (eds), *Models and Idealizations in Science: Artifactual and Fictional Approaches*, 51–70. New York: Springer.

Cartwright, N. (1983). *How the Laws of Physics Lie*. Oxford: Oxford University Press.

Cartwright, N. (1995). "'Ceteris paribus' laws and socio-economic machines". *The Monist* 78(3): 276–94.

Cartwright, N. (2002). "The limits of causal order, from physics to economics". In U. Mäki (ed.), *Fact and Fiction in Economics*, 137–51. Cambridge: Cambridge University Press.

Cartwright, N. (2007). *Hunting Causes and Using Them: Approaches in Philosophy and Economics*. Cambridge: Cambridge University Press.

Carvajalino, J. (2019). "The young Paul Samuelson: mathematics as a language, the operational attitude, and systems in equilibrium". In R. Cord, R. Anderson & W. Barnett (eds), *Paul Samuelson: Master of Modern Economics*, 69–92. London: Springer.

Cassel, G. (1967 [1918]). *The Theory of Social Economy*. New York: Augustus M. Kelley.

Cassini, A. (2021). "Deidealized models". In A. Cassini & J. Redmond (eds), *Models and Idealizations in Science: Artifactual and Fictional Approaches*, 87–114. New York: Springer.

Cauchy, A.-L. (2009 [1821]). *Cauchy's Cours d'Analyse: An Annotated Translation*. Trans. R. Bradley & E. Sandifer. New York: Springer.

Chaitin, G. (1995). "Randomness in arithmetic and the decline and fall of reductionism in pure mathematics". In J. Casti & A. Karlqvist (eds), *Cooperation and Conflict in General Evolutionary Processes*, 89–112. New York: John Wiley.

Chipman, J. (2010). "The Paretian heritage". Corrected version of the original appearing in the *Revue Européene des Sciences Sociales et Cahiers Vilfredo Pareto*, 1970, University of Minnesota, Department of Economics.

Choi, T. (2016). "Physics disarmed: probabilistic knowledge in the works of James Clark Maxwell and George Eliot". In C. Lehleiter (ed.), *Fact and Fiction: Literary and Scientific Cultures in Germany and Britain*, 130–52. Toronto: University of Toronto Press.

Cialowicz, B. & A. Malawski (2011). "The role of banks in the Schumpeterian innovative evolution: an axiomatic set-up". In A. Pyka & M. da Graça Derengowski Fonseca (eds), *Catching Up, Spillovers and Innovation Networks in a Schumpeterian Perspective*, 31–58. London: Springer.

Cirillo, R. (1979). *The Economics of Vilfredo Pareto*. Abingdon: Frank Cass.

Clark, C. (1992). *Economic Theory and Natural Philosophy*. Cheltenham: Edward Elgar.

Clower, R. (1995). "Axiomatics in economics". *Southern Economic Journal* 62(2): 307–19.

Clower, R. (1998). "Three centuries of demand and supply". *Journal of the History of Economic Thought* 20(4): 397–409.

Coase, R. (1978). "Economics and contiguous disciplines". *Journal of Legal Studies* 7(2): 201–11.

Coats, A. (1973). "The economic and social context of the marginal revolution of the 1870's". In R. Black, A. Coats & C. Goodwin (eds), *The Marginal Revolution in Economics: Interpretation and Evaluation*, 37–58. Durham, NC: Duke University Press.

Coats, A. (1976). "Economics and psychology: the death and resurrection of a research programme". In S. Latsis (ed.), *Method and Appraisal in Economics*, 43–64. Cambridge: Cambridge University Press.

Coats, A. (1996). "Utilitarianism, Oxford idealism and Cambridge economics". In P. Groenewegen (ed.), *Economics and Ethics?*, 80–102. London: Routledge.

Coats, A. (2014). *The Historiography of Economics: British and American Economic Essays*. Compiled and edited by R. Backhouse & B. Caldwell. Abingdon: Routledge.

Coddington, A. (1975). "The rationale of general equilibrium theory". *Economic Inquiry* 13(4): 539–58.

Colander, D. (2001). *The Lost Art of Economics: Essays on Economics and the Economics Profession*. Cheltenham: Edward Elgar.

Colander, D. (2009). "What was 'it' that Robbins was defining?" *Journal of the History of Economic Thought* 31(4): 437–48.

Comim, F. (2004). "*The Common Sense of Political Economy* of Philip Wicksteed". *History of Political Economy* 36(3): 475–95.

Conrad, A. & J. Meyer (1958). "The economics of slavery in the ante bellum South". *Journal of Political Economy* 66(2): 95–130.

Cooter, R. (2000). *The Strategic Constitution*. Princeton, NJ: Princeton University Press.

Cooter, R. & T. Ulen (2013). *Law and Economics*. Sixth edn. New York: Pearson.

Corry, L. (1989). "Linearity and reflexivity in the growth of mathematical knowledge". *Science in Context* 3(2): 409–40.

Corry, L. (1993). "Kuhnian issues, scientific revolutions and the history of mathematics". *Studies in History and Philosophy of Science, Part A* 24(1): 95–117.

Corry, L. (2004). *Modern Algebra and the Rise of Mathematical Structures.* Second edn. Basel: Springer.

Corry, L. (2007). "The origin of Hilbert's axiomatic method". In J. Renn (ed.), *The Genesis of General Relativity: Sources and Interpretation, Volume 4 – Theories of Gravitation in the Twilight of Classical Physics: The Promise of Mathematics and the Dream of a Unified Theory,* 758–855. New York: Springer.

Corry, L. (2008). "The development of the idea of proof". In T. Gowers, J. Barrow-Green & I. Leader (eds), *The Princeton Companion to Mathematics,* 129–42. Princeton, NJ: Princeton University Press.

Creedy, J. (1998). *The History of Economic Analysis: Selected Essays.* Cheltenham: Edward Elgar.

Crespo, R. (2013). "Two conceptions of economics and maximisation". *Cambridge Journal of Economics* 37(4): 759–74.

Currie, M. & I. Steedman (1990). *Wrestling with Time: Problems in Economic Theory.* Manchester: Manchester University Press.

Dardo, M. (2004). *Nobel Laureates and Twentieth-Century Physics.* Cambridge: Cambridge University Press.

Darrigol, O. (2012). *A History of Optics: From Greek Antiquity to the Nineteenth Century.* Oxford: Oxford University Press.

Daston, L. (1978). "British responses to psycho-physiology, 1860–1900". *Isis* 69(2): 192–208.

Davies, D. (2007). "Thought experiments and fictional narratives". *Croatian Journal of Philosophy* 7(1): 29–45.

Davis, J. (1989). "Axiomatic general equilibrium theory and referentiality". *Journal of Post Keynesian Economics* 11(3): 424–38.

Davis, J. (2005). "Robbins, textbooks, and the extreme value neutrality view". *History of Political Economy* 37(2): 191–6.

Davis, J. (2011). *Individuals and Identity in Economics.* Cambridge: Cambridge University Press.

Davis, J. (2012). "Mäki on economics imperialism". In A. Lehtinen, J. Kuorikoski & P. Ylikoski (eds), *Economics for Real: Uskali Mäki and the Place of Truth in Economics,* 203–19. Abingdon: Routledge.

Davis, J. (2013). *The Theory of the Individual in Economics: Identity and Value.* Abingdon: Routledge.

Davis, J. (2016). "Economics imperialism versus multidisciplinarity". *History of Economic Ideas* 24(3): 77–94.

Davis, O., M. Hinich & P. Ordeshook (1970). "An expository development of a mathematical model of the electoral process". *American Political Science Review* 64(2): 426–48.

Davis, P. & R. Hersh (1986). *Descartes' Dream: The World According to Mathematics.* Boston, MA: Houghton Mifflin.

Dawson, J. (1984). "Discussion on the foundation of mathematics". *History and Philosophy of Logic* 5(1): 111–29.

Dawson, J. (1986). "Gödel 1931a: Introductory Note to 1931a, 1932e, f and g". In S. Feferman *et al.* (eds), *Kurt Gödel: Collected Works – Volume 1, Publications 1929–1936,* 196–9. New York: Oxford University Press.

REFERENCES

Dawson, J. (2008). "Kurt Gödel". In T. Gowers, J. Barrow-Green & I. Leader (eds), *Princeton Companion to Mathematics*, 819. Princeton, NJ: Princeton University Press.

De Marchi, N. (2002). "Putting evidence in its place: John Mill's early struggles with 'facts in the concrete'". In U. Mäki (ed.), *Fact and Fiction in Economics: Models, Realism and Social Construction*, 304–26. Cambridge: Cambridge University Press.

De Vroey, M. (1975). "The transition from classical to neoclassical economics: a scientific revolution". *Journal of Economic Issues* 9(3): 415–39.

Deane, P. (2001). *The Life and Times of J. Neville Keynes: A Beacon in the Tempest*. Cheltenham: Edward Elgar.

Debreu, G. (1952). "A social equilibrium existence theorem". *Proceedings of the National Academy of Science* 38(10): 886–93.

Debreu, G. (1959). *Theory of Value: An Axiomatic Analysis of Economic Equilibrium*. New York: John Wiley.

Debreu, G. (1974). "Excess demand functions". *Journal of Mathematical Economics* 1(1): 15–21.

Debreu, G. (1984). "Economic theory in the mathematical mode". *American Economic Review* 74(3): 267–78.

Debreu, G. (1986). "Theoretic models: mathematical form and economic content". *Econometrica* 54(6): 1259–70.

Debreu, G. (1987). "Oral history II: an interview". In G. Feiwel (ed.), *Arrow and the Ascent of Modern Economic Theory*, 243–57. London: Macmillan.

Debreu, G. (1989). "Existence of general equilibrium". In J. Eatwell, M. Milgate & P. Newman (eds), *The New Palgrave: General Equilibrium*, 131–8. London: Macmillan.

Debreu, G. (1991a). "The mathematization of economic theory". *American Economic Review* 81(1): 1–7.

Debreu, G. (1991b). "Random walk and life philosophy". *American Economist* 35(2): 3–7.

DeCanio, S. (2014). *Limits of Economic and Social Knowledge*. Basingstoke: Palgrave Macmillan.

Demsetz, H. (1997). "The primacy of economics: an explanation of the comparative success of economics in the social sciences". *Economic Inquiry* 35(1): 1–11.

Detlefsen, M. (1986). *Hilbert's Program: An Essay on Mathematical Instrumentalism*. Dordrecht: Reidel.

Detlefsen, M. (1990). "On an alleged refutation of Hilbert's program using Gödel's first incompleteness theorem". *Journal of Philosophical Logic* 19(4): 343–77.

Dharma-wardana, C. (2013). *A Physicist's View of Matter and Mind*. Singapore: World Scientific.

Diebolt, C. & M. Haupert (eds) (2015). *Handbook of Cliometrics*. Berlin: Springer.

Dierker, E. (1974). *Topological Methods in Walrasian Economics*. Berlin: Springer.

Dieudonné, J. (1970). "The work of Nicolas Bourbaki". *American Mathematical Monthly* 77(?): 134–45.

Dieudonné, J. (1992). *The Music of Reason*. London: Springer.

Dimand, R. & M. Dimand (2002). "Von Neumann and Morgenstern in historical perspective". In C. Schmidt (ed.), *Game Theory and Economic Analysis: A Quiet Revolution in Economics*, 15–32. London: Routledge.

Dobb, M. (1933). "Economic theory and the problems of a socialist economy". *Economic Journal* 43(172): 588–98.

Dobb, M. (1973). *Theories of Value and Distribution since Adam Smith: Ideology and Economic Theory*. Cambridge: Cambridge University Press.

Dobson, P. (2014). *A Chaos of Delight: Science, Religion and Myth and the Shaping of Western Thought*. Abingdon: Routledge.

Dominique, C.-R. (2001). *Market Economies and Natural Laws*. London: Praeger.

Donohue, J. (ed.) (2013). *Law and Economics of Discrimination*. Cheltenham: Edward Elgar.

Dopfer, K. (1989). "Causality and consciousness in economics: concepts of change in orthodox and heterodox economics". In M. Tool & W. Samuels (eds), *Methodology of Economic Thought*. Second edn, 50–64. New Brunswick, NJ: Transaction.

Dornbusch, R. & J. Poterba (eds) (1991). *Global Warming: Economic Policy Responses*. Cambridge, MA: MIT Press.

Downs, A. (1957). *An Economic Theory of Democracy*. New York: Harper.

Doyle, J. (2002). *Extending Mechanics to Minds: The Mechanical Foundations of Psychology and Economics*. Cambridge: Cambridge University Press.

Drakopoulos, S. (2011). "Wicksteed, Robbins and the emergence of mainstream economic methodology". *Review of Political Economy* 23(3): 461–70.

Drakopoulos, S. (2012). "The history of attitudes towards interdependent preferences". *Journal of the History of Economic Thought* 34(4): 541–57.

Dransfield, R. & D. Dransfield (2003). *Key Ideas in Economics*. Cheltenham: Nelson Thomas.

Drescher, S. & S. Engerman (eds) (1998). *A Historical Guide to World Slavery*. Oxford: Oxford University Press.

Duncan, O. (1984). *Notes on Social Measurement: Historical and Critical*. New York: Russell Sage Foundation.

Dunham, W. (1990). *Journey Through Genius: The Great Theories of Mathematics*. London: Penguin.

Dunleavy, P. (2013). *Democracy, Bureaucracy and Public Choice: Economic Explanations in Political Science*. Republished edn. Abingdon: Routledge.

Düppe, T. (2011). *The Making of the Economy: A Phenomenology of Economic Science*. Lanham, MA: Lexington Books.

Düppe, T. (2012a). "Gerard Debreu's secrecy: his life in order and silence". *History of Political Economy* 44(3): 413–49.

Düppe, T. (2012b). "Arrow and Debreu de-homogenized". *Journal of the History of Economic Thought* 34(4): 491–514.

Düppe, T. & R. Weintraub (2014). *Finding Equilibrium: Arrow, Debreu, McKenzie and the Problem of Scientific Credit*. Princeton, NJ: Princeton University Press.

Eboli, F., R. Parrado & R. Roson (2010). "Climate-change feedback on economic growth: explorations with a dynamic general equilibrium model". *Environment and Development Economics* 15(5): 515–33.

Edgeworth, F. (2003 [1881]). *Mathematical Psychics and Further Papers on Political Economy*. Edited by P. Newman. Oxford: Oxford University Press.

Eilenberg, S. & S. Mac Lane (1945). "Relations between homology and homotopy groups of spaces". *Annals of Mathematics* 46(3): 480–509.

Eilenberg, S. & S. Mac Lane (1950). "Relations between homology and homotopy groups of spaces II". *Annals of Mathematics* 51(3): 514–33.

Eilenberg, S. & N. Steenrod (1952). *Foundations of Algebraic Topology*. Princeton, NJ: Princeton University Press.

Elkana, Y. (1986). "The emergence of second-order thinking in classical Greece". In S. Eisenstadt (ed.), *Origins and Diversity of Axial Age Civilizations*, 40–64. Albany, NY: SUNY Press.

Endres, A. & M. Donoghue (2010). "Defending Marshall's 'masterpiece': Ralph Souter's critique of Robbins' *Essay*". *Cambridge Journal of Economics* 34(3): 547–68.

Erickson, P. (2006). *The Nature of Infinitesimals*. New York: Xlibris.

Euclid (1956). *The Thirteen Books of the Elements: Volume I (Books I and II)*. Trans. with introduction and commentary by Sir Thomas Heath. Second edn, unabridged. Mineola, NY: Dover.

Feferman, A. & S. Feferman (2004). *Alfred Tarski: Life and Logic*. Cambridge: Cambridge University Press.

Feferman, S. (1988). "Hilbert's program relativized: proof-theoretical and foundational reductions". *Journal of Symbolic Logic* 53(2): 364–84.

Feferman, S. (2011). "*Lieber Herr Bernays! Lieber Herr Gödel!* Gödel on finitism, constructivity, and Hilbert's program". In M. Baaz *et al.* (eds), *Kurt Gödel and the Foundations of Mathematics: Horizons of Truth*, 111–33. Cambridge: Cambridge University Press.

Feiwel, G. (1982). "Samuelson and contemporary economics: an introduction". In G. Feiwel (ed.), *Samuelson and Neoclassical Economics*, 1–28. Boston, MA: Kluwer Nijhoff.

Ferreirós, J. (1999). *Labyrinth of Thought: A History of Set Theory and Its Role in Modern Mathematics*. New York: Springer.

Ferreirós, J. (2007a). "Ο θεος αριθμητιζει: the rise of pure mathematics as arithmetic with Gauss". In C. Goldstein, N. Schappacher & J. Schwermer (eds), *The Shaping of Arithmetic after C. F. Gauss's Disquisitiones Arithmeticae*, 235–67. Berlin: Springer.

Ferreirós, J. (2007b). *Labyrinth of Thought: A History of Set Theory and Its Role in Modern Mathematics*. Second revised edn. Boston, MA: Birkhäuser.

Fichte, J. (1987 [1800]). *The Vocation of Man*. Trans. P. Preuss. Indianapolis, IN: Hackett.

Fine, B. (2000). "Economics imperialism and intellectual progress: the present as history of economic thought?". *History of Economic Ideas* 32(1): 10–35.

Fine, B. (2004). "Economics imperialism as Kuhnian revolution?". In P. Arestis & M. Sawyer (eds), *The Rise of the Market: Critical Essays on the Political Economy of Neo-Liberalism*, 107–44. Cheltenham: Edward Elgar.

Fine, B. & D. Milonakis (2009). *From Economics Imperialism to Freakonomics: The Shifting Boundaries between Economics and Other Social Sciences*. London: Routledge.

Fischer, S. (1987). "Samuelson, Paul Anthony". In J. Eatwell, M. Milgate & P. Newman (eds), *The New Palgrave: A Dictionary of Economics, Volume III*, 234–41. London: Macmillan.

Fisher, I. (1892). *Mathematical Investigations in the Theory of Value and Price*. New York: Macmillan.

Flatau, P. (2004). "Jevons's one great disciple: Wicksteed and the Jevonian revolution in the second generation". *History of Economics Review* 40(Summer): 69–107.

Flood, R. (2014). "Introduction". In R. Flood, M. McCartney & A. Whitaker (eds), *James Clerk Maxwell: Perspectives on His Life and Work*, 3–16. Oxford: Oxford University Press.

Fogel, R. (1964). *Railroads and American Economic Growth: Essays in Econometric History*. Baltimore, MA: Johns Hopkins University Press.

Fogel, R. & S. Engerman (1974). *Time on the Cross: The Economics of American Negro Slavery.* Boston, MA: Little, Brown.

Forbes, N. & B. Mahon (2014). *Faraday, Maxwell, and the Electromagnetic Field: How Two Men Revolutionized Physics.* Amherst, NY: Prometheus.

Fourier, J. (2007 [1822]). *The Analytical Theory of Heat.* Trans. A. Freeman. New York: Cosimo.

Foxwell, H. (1887). "The economic movement in England". *Quarterly Journal of Economics* 2(1): 84–103.

Fraassen, B. van (1980). *The Scientific Image.* Oxford: Clarendon Press.

Frankhauser, S. (1995). *Valuing Climate Change: The Economics of the Greenhouse.* New York: Earthscan.

Fraser, C. (2003). "The calculus of variations: a historical survey". In H. Jahnke (ed.), *A History of Analysis*, 355–84. London: American Mathematical Society.

Fraser, L. (1932). "How do we want economists to behave?". *Economic Journal* 42(168): 555–70.

Fré, P. (2018). *A Conceptual History of Space and Symmetry: From Plato to the Superworld.* New York: Springer.

Freeman, A., V. Chick & S. Kayatekin (2014). "Samuelson's ghosts: whig history and the reinterpretation of economic theory". *Cambridge Journal of Economics* 38(3): 519–29.

Frege, G. (1950 [1884]). *The Foundations of Arithmetic: A Logico-Mathematical Enquiry into the Concept of Number.* Trans. J. Austin. Oxford: Blackwell.

Frenkiel, É. (2015). *Conditional Democracy: The Contemporary Debate on Political Reform in Chinese Universities.* Colchester: ECPR Press.

Frey, B. & S. Benz (2004). "From imperialism to inspiration: a survey of economics and psychology". In J. Davis, A. Marciano & J. Runde (eds), *Elgar Companion to Economics and Philosophy*, 61–83. Cheltenham: Edward Elgar.

Frigg, R. (2010). "Models and fiction". *Synthese* 172(2): 251–68.

Frisch, R. (1933). "Propagation problems and impulse problems in dynamic economics". In J. Åkerman *et al.* (eds) (1967 [1933]), *Economic Essays in Honour of Gustav Cassel, October 20ʰ 1933*, 171–205. London: Frank Cass.

Frost, G. (2009). "Nobel-winning economist Paul A. Samuelson dies at age 94". MIT News. http://news.mit.edu/2009/obit-samuelson-1213.

Fusfeld, D. (2002). *The Age of the Economist.* Ninth edn. Boston, MA: Addison Wesley.

Galbács, P. (2020). *The Friedman–Lucas Transition in Macroeconomics: A Structuralist Approach.* London: Academic Press.

Gale, D. (1982). *Money: In Equilibrium.* Cambridge: Cambridge University Press.

Gallagher, C. (2006). *The Body Economic: Life, Death, and Sensation in Political Economy and the Victorian Novel.* Princeton, NJ: Princeton University Press.

Gamwell, L. (2016). *Mathematics and Art: A Cultural History.* Princeton, NJ: Princeton University Press.

Gandon, S. (2016). "Wiener and Carnap: a missed opportunity?". In C. Damböck (ed.), *Influences on the Aufbau*, 31–50. New York: Springer.

Gass, S. & A. Assad (2005). *An Annotated Timeline of Operations Research: An Informal History.* Boston, MA: Kluwer.

Gaukroger, S. (2010). *The Collapse of Mechanism and the Rise of Sensibility: Science and the Shaping of Modernity, 1680–1760.* Oxford: Oxford University Press.

Gauthier, Y. (1994). "Hilbert and the internal logic of mathematics". *Synthese* 101(1): 1–14.

Gelfert, A. (2011). "Mathematical formalisms in scientific practice: from denotation to model-based representation". *Studies in History and Philosophy of Science, Part A* 42(2): 272–86.

Georgescu-Roegen, N. (1966). *Analytical Economics.* Cambridge: Cambridge University Press.

Giaquinto, M. (1983). "Hilbert's philosophy of mathematics". *British Journal of the Philosophy of Science* 34(2): 119–32.

Gibbard, A. & H. Varian (1978). "Economic models". *Journal of Philosophy* 75(11): 664–77.

Gibbs, W. (2014 [1902]). *Elementary Principles in Statistical Mechanics.* Mineola, NY: Dover.

Giere, R. (1988). *Explaining Science: A Cognitive Approach.* Chicago, IL: University of Chicago Press.

Gillies, D. (2004). "Can mathematics be used successfully in economics?". In E. Fullbrook (ed.), *A Guide to What's Wrong With Economics*, 187–97. London: Anthem.

Giocoli, N. (2003). *Modelling Rational Agents: From Interwar Economics to Early Modern Game Theory.* Cheltenham: Edward Elgar.

Glock, H.-J. (2003). *Quine and Davidson on Language, Thought and Reality.* Cambridge: Cambridge University Press.

Glymour, C. (1980). "Explanations, tests, unity and necessity". *Noûs* 14(1): 31–50.

Gödel, K. (1931). "Diskussion zur Grundlegung der Mathematik am Sonntag, dem 7. Sept. 1930". *Erkenntnis* 2: 135–51.

Gödel, K. (2003 [1970]). "Letter to Yossef Balas, May 27 1970". In S. Feferman *et al.* (eds), *Kurt Gödel: Collected Works – Volume IV: Correspondence A–G*, 9–11. Oxford: Clarendon Press.

Gödel, K. (2012 [1931]). *On Formally Undecidable Propositions of Principia Mathematica and Related Systems.* Trans B. Melzer & R. Braithwaite. North Chelmsford, MA: Courier.

Godfrey-Smith, P. (2006). "The strategy of model-based science". *Biology and Philosophy* 21(5): 725–40.

Goldstein, R. (2005). *Incompleteness: The Proof and Paradox of Kurt Gödel.* New York: Norton.

Goldstern, M. & H. Judah (1995). *The Incompleteness Phenomenon: A New Course in Mathematical Logic.* London: CRC Press.

Goodman, N. (1991). "Modernizing the philosophy of mathematics". *Synthese* 88(2): 119–26.

Gossen, H. (1854). *Entwickelung der Gesetze des Menschlichen Verkehrs, und der daraus fließenden Regeln für Menschliches Handeln.* Braunschweig: Druck und Verlag von Friedrich Vieweg und Sohn.

Grabiner, J. (1981). "Changing attitudes toward mathematical rigor: Lagrange and analysis in the eighteenth and nineteenth centuries". In H. Jahnke & M. Otte (eds), *Epistemological and Social Problems of the Sciences in the Early Nineteenth Century*, 311–30. Dordrecht: Reidel.

Grabiner, J. (1983). "Who gave you epsilon? Cauchy and the origins of rigorous calculus". *American Mathematical Monthly* 90(3): 185–94.

Grabiner, J. (1984). "Cauchy and Bolzano: tradition and transformation in the history of mathematics". In E. Mendelsohn (ed.), *Transformation and Tradition in the Sciences: Essays in Honor of L. Bernard Cohen*, 105–24. Cambridge: Cambridge University Press.

Grabiner, J. (1996). "The calculus as algebra, the calculus as geometry: Lagrange, Maclaurin, and their legacy". In R. Calinger (ed.), *Vita Mathematica: Historical Research and Integration with Teaching*, 131–44. Washington, DC: Mathematical Association of America.

Grabiner, J. (2010). *A Historian Looks Back: The Calculus as Algebra and Selected Writings.* Washington, DC: Mathematical Association of America.

Gramm, W. (1988). "The movement from real to abstract value theory, 1817–1959". *Cambridge Journal of Economics* 12(2): 225–46.

Grattan-Guinness, I. (2000). *The Search for Mathematical Roots, 1870–1940: Logics, Set Theories and the Foundations of Mathematics from Cantor through Russell to Gödel.* Princeton, NJ: Princeton University Press.

Gratton-Guinness, I. (2002). "'In some parts rather rough': a recently discovered manuscript version of William Stanley Jevons's 'General Mathematical Theory of Political Economy' (1862)". *History of Political Economy* 34(4): 685–726.

Gray, J. (2000). *The Hilbert Challenge.* Oxford: Oxford University Press.

Greasley, D. & L. Oxley (eds) (2011). *Economics and History: Surveys in Cliometrics.* Chichester: Wiley.

Grimmer-Solem, E. & R. Romani (1999). "In search of full empirical reality: historical political economy, 1870–1900". *European Journal of the History of Economic Thought* 6(3): 333–64.

Groenewegen, P. (2003). *Classics and Moderns in Economics, Volume I: Essays on Nineteenth- and Twentieth-Century Economic Thought.* London: Routledge.

Groenewegen, P. (2007). "English marginalism: Jevons, Marshall, and Pigou". In W. Samuels, J. Biddle & J. Davis (eds), *A Companion to the History of Economic Thought.* Second edn, 246–61.

Grofman, B. (ed.) (1993). *Information, Participation, and Choice: An Economic Theory of Democracy in Perspective.* Ann Arbor, MI: University of Michigan Press.

Grosholz, E. (2016). *Starry Reckoning: Reference and Analysis in Mathematics and Cosmology.* New York: Springer.

Gross, M. & V. Tarascio (1998). "Pareto's theory of choice". *History of Political Economy* 30(2): 171–87.

Grüne-Yanoff, T. (2009). "Learning from minimal economic models". *Erkenntnis* 70(1): 81–99.

Grüne-Yanoff, T. & P. Schweinzer (2008). "The roles of stories in applying game theory". *Journal of Economic Methodology* 15(2): 131–46.

Guedj, D. (1985). "Nicholas Bourbaki, collective mathematician: an interview with Claude Chevalley". Trans. J. Gray. *Mathematical Intelligencer* 7(2): 18–22.

Guicciardini, N. (1989). *The Development of Newtonian Calculus in Britain 1700–1800.* Cambridge: Cambridge University Press.

Gunn, D. (2013). *Pneumatology of Matter: A Philosophical Inquiry into the Origins and Meaning of Modern Physical Theory.* Winchester: iff Books.

Haberler, G. von (1927). *Der Sinn der Indexzahlen.* Tübingen: J. C. B. Mohr.

Hacking, I. (1990). *The Taming of Chance.* Cambridge: Cambridge University Press.

Hadamard, J. (1996 [1905]). "Letter from Hadamard to Borel". In W. Ewald (ed.), *From Kant to Hilbert: A Source Book in the Foundations of Mathematics – Volume II*, 1084–5. Oxford: Clarendon Press.

Hahn, F. (1965). "On some problems of proving the existence of an equilibrium in a monetary economy". In F. Hahn & F. Brechling (eds), *The Theory of Interest Rates*, 126–35. London: Macmillan.

Hahn, F. (1973a). "The winter of our discontent". *Economica* 40(3): 322–30.

Hahn, F. (1973b). *On the Notion of Equilibrium in Economics*. Cambridge: Cambridge University Press.

Hahn, F. (1982). "Stability". In K. Arrow & M. Intriligator (eds), *Handbook of Mathematical Economics: Volume I*, 745–93. Amsterdam: North-Holland.

Hall, M. (1837). *Memoirs on the Nervous System*. London: Sherwood, Gilber & Piper.

Hallett, M. (1994). "Hilbert's axiomatic method and the laws of thought". In A. George (ed.), *Mathematics and Mind*, 158–200. Oxford: Oxford University Press.

Halpin, H. (2013). *Social Semantics: The Search for Meaning on the Web*. New York: Springer.

Händler, E. (1980). "The logical structure of modern neoclassical static microeconomic equilibrium theory". *Erkenntnis* 15(1): 33–53.

Hands, W. (2001). *Reflection Without Rules: Economic Methodology and Contemporary Science Theory*. Cambridge: Cambridge University Press.

Hands, W. (2009a). "Did Milton Friedman's positive methodology license the formalist revolution?". In U. Mäki (ed.), *The Methodology of Positive Economics: Reflections on the Milton Friedman Legacy*, 143–64. Cambridge: Cambridge University Press.

Hands, W. (2009b). "Effective tension in Robbins' economic methodology". *Economica* 76(Supp.): 831–44.

Hands, W. (2012). "What a difference a sum (Σ) makes: success and failure in the rationalization of demand". *Journal of the History of Economic Thought* 34(3): 379–96.

Hands, W. (2016). "Derivational robustness, credible substitute systems and mathematical economic models: the case of stability analysis in Walrasian general equilibrium theory". *European Journal for Philosophy of Science* 6(1): 31–53.

Hands, W. (2019). "Re-examining Samuelson's operationalist methodology". In R. Cord, R. Anderson & W. Barnett (eds), *Paul Samuelson: Master of Modern Economics*, 39–68. Basingstoke: Palgrave Macmillan.

Hargittai, B. & I. Hargittai (2016). *Wisdom of the Martians of Science: In their Own Words with Commentaries*. Singapore: World Scientific.

Harrod, R. (1938). "Scope and method of economics". *Economic Journal* 48(191): 383–412.

Hausman, D. (1994). "Kuhn, Lakatos and the character of economics". In R. Backhouse (ed.), *New Directions in Economic Methodology*, 197–217. London: Routledge.

Hausman, D. (2008). "Mindless or mindful economics: a methodological evaluation". In A. Caplin & A. Schotter (eds), *The Foundations of Positive and Normative Economics: A Handbook*, 125–51. Oxford: Oxford University Press.

Hawtrey, R. (1926). *The Economic Problem*. London: Longmans.

Hay, C. (2002). *Political Analysis: A Critical Introduction*. Basingstoke: Palgrave Macmillan.

Hay, C. (2004). "Theory, stylized heuristic or self-fulfilling prophecy? The status of rational choice theory in public administration". *Public Administration* 82(1): 39–62.

Heaviside, O. (1888). "On electromagnetic waves, especially in relation to the vorticity of the impressed forces; and the forced vibrations of electromagnetic systems". *The London, Edinburgh, and Dublin Philosophical Magazine and Journal of Science* 25(153): 130–56.

Henderson, A. (2018). *Algebraic Art: Mathematical Formalism and Victorian Culture*. Oxford: Oxford University Press.

Henderson, J. (1994). "The place of economics in the hierarchy of the sciences: section F from Whewell to Edgeworth". In P. Mirowski (ed.), *Natural Images in Economic Thought: "Markets Read in Tooth and Claw"*, 484–535. Cambridge: Cambridge University Press.

Henderson, L. (2013). *The Fourth Dimension and Non-Euclidean Geometry in Modern Art.* Revised edn. Cambridge, MA: MIT Press.

Hertz, H. (1896). *Miscellaneous Papers.* Trans. D. Jones & G. Schott. London: Macmillan.

Hesse, M. (1966). *Models and Analogies in Science.* Notre Dame, IN: Notre Dame University Press.

Hicks, J. (1934). "A reconsideration of the theory of value: part I". *Economica* 1(1): 52–76.

Hicks, J. (1939). *Value and Capital.* Oxford: Clarendon Press.

Hicks, J. (1984). "The formation of an economist". In D. Helm (ed.), *The Economics of John Hicks*, 281–90. Oxford: Blackwell.

Hilbert, D. (1890). "Ueber die Theorie der Algebrischen Formen". *Mathematische Annalen* 36(4): 473–534.

Hilbert, D. (1893). "Ueber die vollen Invariantensysteme". *Mathematische Annalen* 42(3): 313–73.

Hilbert, D. (1902). "Mathematical problems: lecture delivered before the International Congress of Mathematicians at Paris in 1900, translated by Mary Winston Newson". *Bulletin of the American Mathematical Society* 8(10): 437–79.

Hilbert, D. (1905). "On the foundations of logic and arithmetic". Trans. G. Halstead. *The Monist* 15(3): 338–52.

Hilbert, D. (1924). *Methoden der Mathematischen Physik.* Berlin: Springer.

Hilbert, D. (1930a). "Probleme der Grundlegung der Mathematik". *Mathematische Annalen* 102(1): 1–9.

Hilbert, D. (1930b). "Naturerkennen und Logik". *Die Naturwissenschaften* 18(47–49): 959–63.

Hilbert, D. (1950 [1899]). *Foundations of Geometry.* Trans. from *Grundlagen der Geometrie* by E. Townsend. Reprint edn. La Salle, IL: Open Court.

Hilbert, D. (1967 [1925]). "On the infinite". Trans. S. Bauer-Mengelberg. In J. van Heijenoort (ed.) (1967), *From Frege to Gödel: A Source Book in Mathematical Logic, 1879–1931*, 367–92. Cambridge, MA: Harvard University Press.

Hilbert, D. (1996 [1922]). "The new grounding of mathematics: first report". In W. Ewald (1996), *From Kant to Hilbert: A Source Book in the Foundations of Mathematics, Volume II*, 1115–34. Oxford: Clarendon Press.

Hilbert, D. (1998 [1897]). *The Theory of Algebraic Number Fields.* Trans. I. Adamson. New York: Springer.

Hilbert, D. & P. Bernays (1934). *Grundlagen der Mathematik, Volumen I.* Berlin: Springer.

Hilbert, D. & P. Bernays (2003 [1934]). *Foundations of Mathematics, Volume 1.* Partial trans. I. Mueller. Bernays Project Text Number 12. www.phil.cmu.edu/projects/bernays/Pdf/bernays12-2_2003-06-25.pdf.

Hilbert, D. & S. Cohn-Vossen (1999 [1932]). *Geometry and Imagination.* Trans. P. Nemenyi. Reprinted second edn. Providence, RI: American Mathematical Society.

Hildenbrand, W. (1983). "Introduction". In G. Debreu, *Mathematical Economics: Twenty Collected Papers of Gérard Debreu*, 1–29. Cambridge: Cambridge University Press.

Hill, R. & T. Myatt (2010). *The Anti-Economics Textbook: A Critical Thinker's Guide to Microeconomics.* London: Zed Books.

Hindriks, F. (2006). "Tractability assumptions and the Musgrave–Mäki typology". *Journal of Economic Methodology* 13(4): 401–23.

Hinich, M. & M. Munger (1997). *Analytical Politics*. Cambridge: Cambridge University Press.

Hirshleifer, J. (1976). *Price Theory and Applications*. New York: Prentice Hall.

Hirshleifer, J. (1985). "The expanding domain of economics". *American Economic Review* 75(6): 53–68.

Hodgson, G. (1994). "Some remarks on 'economic imperialism' and international political economy". *Review of International Political Economy* 1(1): 21–8.

Hodgson, G. (1999). *Economics and Utopia: Why the Learning Economy Is Not the End of History*. London: Routledge.

Hodgson, G. (2001). *How Economics Forgot History: The Problem of Historical Specificity in Social Science*. London: Routledge.

Hodgson, G. (2008). "Marshall, Schumpeter and the shifting boundaries of economics and sociology". In Y. Shionoya & T. Nishizawa (eds), *Marshall and Schumpeter on Evolution: Economic Sociology of Capitalist Development*, 93–115. Cheltenham: Edward Elgar.

Hodgson, G. (2013). "Dr Blaug's diagnosis: is economics sick?". In M. Boumans & M. Klaes (eds), *Mark Blaug: Rebel with Many Causes*, 78–97. Cheltenham: Edward Elgar.

Hoffman, P. (1998). *The Man Who Loved Only Numbers: The Story of Paul Erdos and the Search for Mathematical Truth*. New York: Hyperion.

Hollis, M. (1998). *Trust within Reason*. Cambridge: Cambridge University Press.

Horn, K. (2009). *Roads to Wisdom: Conversations with Ten Nobel Laureates in Economics*. Cheltenham: Edward Elgar.

Houthakker, H. (1950). "Revealed preference and the utility function". *Economica* 17(2): 159–74.

Houthakker, H. (1983). "On consumption theory". In C. Brown & R. Solow (eds), *Paul Samuelson and Modern Economic Theory*, 57–68. New York: McGraw Hill.

Howey, R. (1973). "The origins of marginalism". In R. Black, A. Coats & C. Goodwin (eds), *The Marginal Revolution in Economics: Interpretation and Evaluation*, 15–36. Durham, NC: Duke University Press.

Howson, S. (2004). "The origins of Lionel Robbins's *Essay on the Nature and Significance of Economic Science*". *History of Political Economy* 36(3): 413–43.

Howson, S. (2011). *Lionel Robbins*. Cambridge: Cambridge University Press.

Hurwicz, L. (1987). "Oral history III: an interview with Leonid Hurwicz". In G. Feiwel (ed.), *Arrow and the Ascent of Modern Economic Theory*, 258–91. London: Macmillan.

Hurwicz, L. & M. Richter (1979). "An integrability condition with applications to utility theory and thermodynamics". *Journal of Mathematical Economics* 6(1): 7–14.

Hurwicz, L. & H. Uzawa (1971). "On the problem of integrability in economics". In J. Chipman *et al.* (eds), *Preferences, Utility, and Demand*, 174–213. New York: Harcourt Brace Jovanovich.

Hutchison, T. (1935). "A note on tautologies and the nature of economic theory". *Review of Economic Studies* 2(2): 159–61.

Hutchison, T. (1938). *The Significance and Basic Postulates of Economic Theory*. London: Macmillan.

Hutchison, T. (1953). *A Review of Economic Doctrines, 1870–1929*. Oxford: Clarendon Press.

Hutchison, T. (1982). "The politics and philosophy in Jevons's political economy". *The Manchester School* 50(4): 366–78.

Huxley, T. (1874). "On the hypothesis that animals are automata, and its history". In T. Huxley (2011 [1894]), *Collected Essays: Volume 1, Methods and Results*, 199–250. Cambridge: Cambridge University Press.

Iglowitz, J. (2012). *Consciousness and Reality: Final and Definite Conclusions*. Granite Bay, CA: Jerryspace Publishing.

Ingrao, B. & G. Israel (1990). *The Invisible Hand: Economic Equilibrium in the History of Science*. Trans. I. McGilvray. Cambridge, MA: MIT Press.

Inoue, T. (2012). "Quételet's influence on W. S. Jevons: from subjectivism to objectivism". In Y. Ikeda & K. Yagi (eds), *Subjectivism and Objectivism in the History of Economic Thought*, 48–58. Abingdon: Routledge.

Ireland, P. (2003). "Endogenous money or sticky prices?". *Journal of Monetary Economics* 50(8): 1623–48.

Israel, G. (1981). "'Rigor' and 'axiomatics' in modern mathematics". *Fundamenta Scientiae* 2(2): 205–19.

Israel, G. & A. Gasca (2009). *The World as a Mathematical Game: John von Neumann and Twentieth Century Science*. Trans. I. McGilvray. Basel: Birkhäuser Verlag.

Jacobs, K. (1992). *Invitation to Mathematics*. Princeton, NJ: Princeton University Press.

Jacyna, S. (1981). "The physiology of mind, the unity of nature, and the moral order in Victorian thought". *British Journal for the History of Science* 14(2): 109–32.

Jaffé, W. (1964). "New light on an old quarrel: Barone's unpublished review of Wicksteed's 'Essay on the Coordination of the Laws of Distribution' and related documents". *Cahiers Vilfredo Pareto* 3: 61–102.

Jaffé, W. (1967). "Walras' theory of *tâtonnement*: critique of recent interpretations". *Journal of Political Economy* 75(1): 1–19.

Jaffé, W. (1976). "Menger, Jevons and Walras de-homogenized". *Economic Inquiry* 14(4): 511–24.

Jagnon, R. (2006). "Edmund Husserl on the applicability of formal geometry". In E. Carson & R. Huber (eds), *Intuition and the Axiomatic Method*, 67–85. New York: Springer.

Jahnke, H. (2003). "Algebraic analysis in the 18th century". In H. Jahnke (ed.), *A History of Analysis*, 105–36. Providence, RI: American Mathematical Society.

Jarvis, L. & V. Mosini (2017). "The ubiquity of the notion of equilibrium in biology, and its relation with equilibrium in economics". In V. Mosini (ed.), *Equilibrium in Economics: Scope and Limits*, 60–73. London: Routledge.

Jennings, R. (1969 [1855]). *Natural Elements of Political Economy*. New York: Augustus M. Kelley.

Jesseph, D. (1993). *Berkeley's Philosophy of Mathematics*. Chicago, IL: University of Chicago Press.

Jevons, W. S. (1862). "Notice of a general mathematical theory of political economy". Paper read to the Annual Meeting of the British Association for the Advancement of Science, Cambridge, October 1862.

Jevons, W. S. (1863). "Balance". In H. Watts (ed.), *A Dictionary of Chemistry and the Allied Branches of Other Sciences*, 481–91. London: Longman, Green, Roberts & Green.

Jevons, W. S. (1869). *The Substitution of Similars: The True Principle of Reasoning, Derived from a Modification of Aristotle's Dictum*. London: Macmillan.

Jevons, W. S. (1870). "On the natural laws of muscular exertion". *Nature* 2(June 30): 158–60.

Jevons, W. S. (1876). "The future of political economy". *Fortnightly Review* 20(119): 617–31.

Jevons, W. S. (1877). "John Stuart Mill's philosophy tested: part I". *Contemporary Review* 31(12): 167–82.

Jevons, W. S. (1878). *Political Economy*. London: Macmillan.

Jevons, W. S. (1882). *The State in Relation to Labour*. London: Macmillan.

Jevons, W. S. (1883). "Amusements of the people". In W. S. Jevons, *Methods of Social Reform and Other Papers*. Compiled and edited by H. Jevons, 1–27. London: Macmillan.

Jevons, W. S. (1981). "Extracts from the personal diaries, 1856–60". In R. Black (ed.), *Papers and Correspondence of William Stanley Jevons – Volume VII: Papers in Political Economy*, 113–20. London: Macmillan.

Jevons, W. S. (2010 [1887]). *The Principles of Science: A Treatise on Logic and Scientific Method*. Whitefish, MT: Kessinger Publishing.

Jevons, W. S. (2013 [1871/1879]). *The Theory of Political Economy*. Composite of the first and second editions. Basingstoke: Palgrave Macmillan.

Jevons, W. S. (2014 [1874]). *The Principles of Science: A Treatise on Logic and Scientific Method*. Charleston, SC: Nabu Press.

Jolink, A. (1996). *The Evolutionist Economics of Léon Walras*. London: Routledge.

Jones, T. (2001). "*Unifying Scientific Theories*, by Margaret Morrison". *Mind* 110(4): 1097–102.

Kakutani, S. (1941). "A generalization of Brouwer's fixed point theorem". *Duke Mathematical Journal* 8(3): 457–9.

Kalman, D. (2009). *Uncommon Mathematical Excursions: Polynomia and Related Realms*. Washington, DC: Mathematical Association of America.

Keen, S. (2013). "Predicting the 'Global Financial Crisis': Post-Keynesian macroeconomics". *Economic Record* 89(2): 228–54.

Kelly, J. (1987). "An interview with Kenneth J. Arrow". *Social Choice and Welfare* 4(1): 43–62.

Kemp, G. (2012). *Quine Versus Davidson: Truth, Reference, and Meaning*. Oxford: Oxford University Press.

Keynes, J. N. (1999 [1890/1917]). *The Scope and Method of Political Economy*. Kitchener, ON: Batoche Books.

Kim, M. G. (2003). *Affinity, that Elusive Dream: A Genealogy of the Chemical Revolution*. Cambridge, MA: MIT Press.

Kirman, A. (2011). "Walras's unfortunate legacy". In P. Bridel (ed.), *General Equilibrium Analysis: A Century After Walras*, 109–33. London: Routledge.

Kirman, A. & K.-J. Koch (1986). "Market excess demand functions in exchange economies with identical preferences and collinear endowments". *Review of Economic Studies* 53(3): 457–63.

Kirman, A. & J.-B. Zimmermann (2001). "Introduction". In A. Kirman & J.-B. Zimmermann (eds), *Economics with Heterogeneous Interacting Agents*, 1–9. New York: Springer.

Kitcher, P. (1976). "Hilbert's epistemology". *Philosophy of Science* 43(1): 99–115.

Kitcher, P. (1981). "Explanatory unification". *Philosophy of Science* 48(4): 507–31.

Kitcher, P. (1989). "Explanatory unification and the causal structure of the world". In P. Kitcher & W. Salmon (eds), *Scientific Explanation*, 410–506. Minneapolis, MN: University of Minnesota Press.

Kitching, G. (2012). *Development and Underdevelopment in Historical Perspective: Populism, Nationalism and Industrialization*. Abingdon: Routledge.

Klamer, A. (1994). "Formalism in twentieth century economics". In P. Boettke (ed.), *Elgar Companion to Austrian Economics*, 48–53. Cheltenham: Edward Elgar.

Kline, M. (1953). *Mathematics in Western Culture*. Oxford: Oxford University Press.

Kline, M. (1972). *Mathematical Thought from Ancient to Modern Times: Volume 3*. Oxford: Oxford University Press.

Kneale, W. (1949). *Probability and Induction*. Oxford: Clarendon Press.

Knuuttila, T. (2009). "Isolating representations versus credible constructions? Economic modelling in theory and practice". *Erkenntnis* 70(1): 59–80.

Knuuttila, T. (2011). "Modelling and representing: an artefactual approach to model-based representation". *Studies in History and Philosophy of Science, Part A* 42(2): 262–71.

Koetsier, T. (2002). "Lakatos' mitigated scepticism in the philosophy of mathematics". In G. Kampis, L. Kvasz & M. Stöltzner (eds), *Appraising Lakatos: Mathematics, Methodology, and the Man*, 189–210. Berlin: Springer.

Konüs, A. (1939). "The problem of the true index of the cost of living". *Econometrica* 7(1): 10–29.

Kreisel, D. (2012). *Economic Woman: Demand, Gender, and Narrative Closure in Eliot and Hardy*. Toronto: University of Toronto Press.

Kreisel, G. (1976). "What have we learned from Hilbert's second problem?". In F. Browder (ed.), *Mathematical Developments Arising from Hilbert's Problems*, 93–130. Providence, RI: American Mathematical Society.

Krugman, P. (2009). *The Return of Depression Economics and the Crisis of 2008*. New York: Norton.

Kuorikoski, J. (2021). "There are no mathematical explanations". *Philosophy of Science* 88(2): 189–212.

Kuorikoski, J. & A. Lehtinen (2009). "Incredible worlds, credible results". *Erkenntnis* 70(1): 119–31.

Kuorikoski, J. & A. Lehtinen (2010). "Economics imperialism and solution concepts in political science". *Philosophy of the Social Sciences* 40(3): 347–74.

Kuorikoski, J. & C. Marchionni (2014). "Philosophy of economics". In S. French & J. Saatsi (eds), *Bloomsbury Companion to the Philosophy of Science*, 314–33. London: Bloomsbury.

Kuorikoski, J., A. Lehtinen & C. Marchionni (2010). "Economic modelling as robustness analysis". *British Journal for the Philosophy of Science* 61(3): 541–67.

Kuorikoski, J., A. Lehtinen & C. Marchionni (2012). "Robustness analysis disclaimer: please read the manual before use!". *Biology and Philosophy* 27(6): 891–902.

Kydland, F. & E. Prescott (1982). "Time to build and aggregate fluctuations". *Econometrica* 50(6): 1345–70.

Kydland, F. & E. Prescott (1996). "The computational experiment: an econometric tool". *Journal of Economic Perspectives* 10(1): 69–85.

Lagrange, J.-L. (2009 [1788]). *Mécanique Analytique, 2 Volumes*. Cambridge: Cambridge University Press.

Lakatos, I. (1978). *The Methodology of Scientific Research Programmes: Philosophical Papers Volume 1*. Edited by J. Worrall & G. Currie. Cambridge: Cambridge University Press.

Landes, W. & R. Posner (1993). "The influence of economics on law: a quantitative study". *Journal of Law and Economics* 36(1): 385–424.

Laugwitz, D. (2000). "Controversies about numbers and functions". In E. Grosholz & H. Breger (eds), *The Growth of Mathematical Knowledge*, 177–98. Dordrecht: Kluwer.

Lavoie, D. (1985). *Rivalry and Central Planning: The Socialist Calculation Debate Reconsidered.* Cambridge: Cambridge University Press.

Lawson, T. (2003). *Reorienting Economics: Economics as Social Theory.* London: Routledge.

Laycock, T. (1845). "On the reflex function of the brain". *British and Foreign Medical Review* 19: 298–311.

Lazear, E. (2000). "Economic imperialism". *Quarterly Journal of Economics* 115(1): 99–146.

Leamer, E. (1983). "Let's take the con out of econometrics". *American Economic Review* 73(1): 31–43.

Leapard, D. (2013). "Samuelson, Paul". In D. Dieterle (ed.), *Economic Thinkers: A Biographical Encyclopedia*, 355–7. Santa Barbara, CA: Greenwood.

Lee, M. (2012). *Uncertain Chances: Science, Skepticism, and Belief in Nineteenth-Century American Literature.* Oxford: Oxford University Press.

Lenfant, J.-S. (2012). "Indifference curves and the ordinalist revolution". *History of Political Economy* 44(1): 113–55.

Leonard, R. (1992). "Creating a context for game theory". In R. Weintraub (ed.), *Toward a History of Game Theory*, 29–76. Durham, NC: Duke University Press.

Leonard, R. (2010). *Von Neumann, Morgenstern, and the Creation of Game Theory: From Chess to Social Science, 1900–1960.* Cambridge: Cambridge University Press.

Leslie, T. (1862). "The love of money". In T. Leslie (1969), *Essays in Political Economy*, 1–8. New York: Augustus M. Kelley.

Leslie, T. (2013 [1870]). *Land Systems and Industrial Economy of Ireland, England and Continental Countries.* Lenox, MA: HardPress Publishing.

Lewin, S. (1996). "Economics and psychology: lessons for our own day from the early twentieth century". *Journal of Economic Literature* 34(3): 1293–323.

Lewis, D. (1973). "Causation". *Journal of Philosophy* 70(17): 556–67.

Lipsey, R. (2009). "Some legacies of Robbins' *An Essay on the Nature and Significance of Economic Science*". *Economica* 76(Supp.): 845–56.

Longair, M. (2016). *Maxwell's Enduring Legacy: A Scientific History of the Cavendish Laboratory.* Cambridge: Cambridge University Press.

Lucas, R. (1972). "Expectations and the neutrality of money". *Journal of Economic Theory* 4(2): 103–24.

Lucas, R. (1981). *Studies in Business-Cycle Theory.* Cambridge, MA: MIT Press.

Lucas, R. (1988). "On the mechanics of economic development". *Journal of Monetary Economics* 22(1): 3–42.

Lucas, R. (2001). "Professional memoire". Transcript of lecture delivered in the Nobel Economists Lecture Series, Trinity University, San Antonio, Texas. http://homepage.ntu.edu.tw/~yitingli/file/Workshop/memoir.pdf.

Lundahl, M. & E. Wadensjö (1984). *Unequal Treatment: A Study in the Neo-Classical Theory of Discrimination.* London: Croom Helm.

Lyons, J., L. Cain & S. Williamson (eds) (2008). *Relections on the Cliometrics Revolution: Conversations with Economic Historians*. London: Routledge.

Maas, H. (2005a). *William Stanley Jevons and the Making of Modern Economics*. Cambridge: Cambridge University Press.

Maas, H. (2005b). "Jevons, Mill and the private laboratory of the mind". *The Manchester School* 73(5): 620–49.

Maas, H. (2009). "Disciplining boundaries: Lionel Robbins, Max Weber, and the borderlands of economics, history, and psychology". *Journal of the History of Economic Thought* 31(4): 500–17.

Maas, H. (2014). *Economic Methodolgy: A Historical Introduction*. Trans. L. Waters. Abingdon: Routledge.

Macciò, D. (2015). "G. E. Moore's philosophy and Cambridge economics: Ralph Hawtrey on ethics and methodology". *European Journal of the History of Economic Thought* 22(2): 163–97.

Mainzer, K. (1996). *Symmetries of Nature: A Handbook for Philosophy of Nature and Science*. Second edn. Berlin: De Gruyer.

Mäki, U. (2000). "Kinds of assumptions and their truth: shaking an untwisted F-twist". *Kyklos* 53(3): 317–36.

Mäki, U. (2001a). "Explanatory unification: double and doubtful". *Philosophy of the Social Sciences* 31(4): 488–506.

Mäki, U. (2001b). "The way the world works (www): towards an ontology of theory choice". In U. Mäki (ed.), *The Economic World View: Studies in the Ontology of Economics*, 369–89. Cambridge: Cambridge University Press.

Mäki, U. (2002). "Symposium on explanations and social ontology 2: explanatory ecumenism and economics imperialism". *Economics and Philosophy* 18(2): 235–57.

Mäki, U. (2009a). "Economics imperialism: concepts and constraints". *Philosophy of the Social Sciences* 39(3): 351–80.

Mäki, U. (2009b). "MISSing the world: models as isolations and credible surrogate systems". *Erkenntnis* 70(1): 29–43.

Mandik, P. & J. Weisberg (2008). "Type-Q materialism". In C. Wrenn (ed.), *Naturalism, Reference, and Ontology: Essays in Honor of Roger F. Gibson*, 223–46. New York: Peter Lang.

Mandler, M. (1999). *Dilemmas in Economic Theory: Persisting Foundational Problems of Microeconomics*. Oxford: Oxford University Press.

Mantel, R. (1974). "On the characterization of aggregate excess demand". *Journal of Economic Theory* 7(3): 348–53.

Marshall, A. (2013 [1890/1920]). *Principles of Economics*. Eighth edn. Basingstoke: Palgrave Macmillan.

Martin, A. (2009). "Critical realism and the Austrian paradox". *Cambridge Journal of Economics* 33(3): 517–30.

Martins, N. (2014). *The Cambridge Revival of Political Economy*. London: Routledge.

Masini, F. (2009). "*Economics* and *political economy* in Lionel Robbins's writings". *Journal of the History of Economic Thought* 31(4): 421–36.

Masini, F. (2010). "Economics and laicism: the struggle for a 'humane discipline'". In D. Parisi & S. Solari (eds), *Humanism and Religion in the History of Economic Thought*, 42–58. Rome: Franco Angeli.

Maskin, E. (2001). "Kinds of theory". In T. Negishi, R. Ramachandran & K. Mino (eds), *Economic Theory, Dynamics, and Markets: Essays in Honor of Ryuzo Sato*, 45–56. Norwell, MA: Kluwer.

Mathias, A. (2014). "Hilbert, Bourbaki and the scorning of logic". In C. Chong *et al.* (eds), *Infinity and Truth*, 47–156. Singapore: World Scientific.

Maudsley, H. (1876). *The Physiology of Mind*. London: Macmillan.

Maxwell, J. C. (1856). "On Faraday's lines of force". *Transactions of the Cambridge Philosophical Society* 10(1): 27–83.

Maxwell, J. C. (1861–2). "On physical lines of force". *The London, Edinburgh and Dublin Philosophical Magazine and Journal of Science*, Part I: 21(139): 161–75; Part II: 21(140): 281–91, 21(141): 338–48; Part III: 23(151): 12–24; Part IV: 23(152): 85–95.

Maxwell, J. C. (1865). "A dynamical theory of the electromagnetic field". *Philosophical Transactions of the Royal Society of London* 155: 459–512.

Maxwell, J. C. (1873). *A Treatise on Electricity and Magnetism, 2 Volumes*. London: Macmillan.

Maxwell, J. C. (1890 [1870]). "Address to the mathematical and physical sections of the British Association, Liverpool, September 15, 1870". In W. Niven (ed.), *The Scientific Papers of James Clerk Maxwell, Volume II*, 215–29. Cambridge: Cambridge University Press.

Mayberry, J. (2000). *The Foundations of Mathematics in the Theory of Sets*. Cambridge: Cambridge University Press.

McCann, C. (1994). *Probability Foundations of Economic Theory*. London: Routledge.

McCloskey, D. (1983). "The rhetoric of economics". *Journal of Economic Literature* 21(2): 481–517.

McCloskey, D. (1990). "Their blackboard, right or wrong: a comment on contested exchange". *Politics and Society* 18(2): 223–32.

McCloskey, D. (2005). "Other things equal: Samuelsonian economics". In J. Wood & M. McClure (eds), *Paul A. Samuelson: Critical Assessments of Contemporary Economists, Volume 3*, 74–80. London: Routledge.

McClure, M. (2002). *Pareto, Economics and Society: The Mechanical Analogy*. London Routledge.

McClure, M. (2005). "Equilibrium and Italian fiscal sociology: a reflection on the Pareto–Griziotti and Pareto–Sensini letters on fiscal theory". *European Journal of the History of Economic Thought* 12(4): 609–33.

McClure, M. (2010). "Pareto, Pigou and third-party consumption: divergent approaches to welfare theory with implications for the study of public finance". *European Journal of the History of Economic Thought* 17(4): 635–57.

McKenzie, L. (1954). "On equilibrium in Graham's model of world trade and other competitive systems". *Econometrica* 22(2): 147–61.

McKenzie, L. (1959). "On the existence of general equilibrium for a competitive market". *Econometrica* 27(1): 54–71.

McKenzie, L. (1989). "General equilibrium". In J. Eatwell, M. Milgate & P. Newman (eds), *The New Palgrave: General Equilibrium*, 1–35. London: Macmillan.

McMullin, E. (1985). "Galilean idealization". *Studies in History and Philosophy of Science, Part A* 16(3): 247–73.

Medema, S. (1997). "The trial of *Homo economicus*: what law and economics tells us about the development of economic imperialism". *History of Political Economy* 29(Supp.): 122–42.

Meek, R. (1967). *Economics and Ideology and Other Essays*. London: Chapman & Hall.

Mendelsohn, R. (ed.) (2022). *Climate Change Economics: Commemoration of Nobel Prize for William Nordhaus*. Singapore: World Scientific.

Menger, C. (1981 [1871]). *Principles of Economics*. Trans. J. Dingwall & B. Hoselitz. Auburn, AL: Mises Institute.

Menzler-Trott, E. (2007). *Logic's Lost Genius: The Life of Gerhard Gentzen*. Trans. C. Smoryński & E. Griffor. Providence, RI: American Mathematical Society.

Merzbach, U. & C. Boyer (2011). *A History of Mathematics*. Third edn. Chichester: John Wiley.

Miller, A. (1984). *Imagery in Scientific Thought: Creating 20th-Century Physics*. Boston, MA: Birkhäuser.

Milonakis, D. & B. Fine (2009). *From Political Economy to Economics: Method, the Social and the Historical in the Evolution of Economic Theory*. London: Routledge.

Mirowski, P. (1991). *More Heat than Light: Economics as Social Physics, Physics as Nature's Economics*. Paperback edn. Cambridge: Cambridge University Press.

Mirowski, P. (2002). *Machine Dreams: Economics Becomes a Cyborg Science*. Cambridge: Cambridge University Press.

Mirowski, P. (2012a). "The Cowles Commission as an anti-Keynesian stronghold 1943–54". In P. Duarte & G. Lima (eds), *Microfoundations Reconsidered: The Relationship of Micro and Macroeconomics in Historical Perspective*, 131–67. Cheltenham: Edward Elgar.

Mirowski, P. (2012b). "The unreasonable efficacy of mathematics in modern economics". In U. Mäki (ed.), *Philosophy of Economics*, 159–97. London: Elsevier.

Moore, G. C. G. (1995). "T. E. Cliffe Leslie and the English *Methodenstreit*". *Journal of the History of Economic Thought* 17(1): 57–77.

Moore, G. C. G. (2003). "John Neville Keynes's solution to the English *Methodenstreit*". *Journal of the History of Economic Thought* 23(1): 5–38.

Moore, G. H. (1982). *Zermelo's Axiom of Choice: Its Origins, Development and Influence*. New York: Springer-Verlag.

Morgan, M. (2001). "Models, stories and the economic world". *Journal of Economic Methodology* 8(3): 361–84.

Morgan, M. (2006). "Economic man as model man: ideal types, idealization and caricatures". *Journal of the History of Economic Thought* 28(1): 1–38.

Morgan, M. (2012). *The World in the Model: How Economists Work and Think*. Cambridge: Cambridge University Press.

Morgenstern, O. (1941). "Professor Hicks on value and capital". *Journal of Political Economy* 49(3): 361–93.

Morgenstern, O. (1951). "Abraham Wald, 1902–1950". *Econometrica* 19(4): 361–7.

Morishima, M. (1984). "The good and bad uses of mathematics". In P. Wiles & G. Routh (eds), *Economics in Disarray*, 51–73. Oxford: Blackwell.

Morrison, M. (1994). "Unified theories and disparate things". *Proceedings of the Biennial Meeting of the Philosophy of Science Association* 1994(2): 365–73.

Morrison, M. (1999). "Models as autonomous agents". In M. Morgan & M. Morrison (eds), *Models as Mediators: Perspectives on Natural and Social Science*, 38–65. Cambridge: Cambridge University Press.

Morrison, M. (2000). *Unifying Scientific Theories: Physical Concepts and Mathematical Structures*. Cambridge: Cambridge University Press.

Morrison, M. (2018). "Building theories: strategies not blueprints". In D. Danks & E. Ippoliti (eds), *Building Theories: Heuristics and Hypotheses in Sciences*, 21–44. New York: Springer.

Moscati, I. (2013). "Were Jevons, Menger, and Walras really cardinalists? On the notion of measurement in utility theory, psychology, mathematics, and other disciplines, 1870–1910". *History of Political Economy* 45(3): 373–414.

Mosselmans, B. (2001). "Jevons, William Stanley (1835–1882)". In J. Powell (ed.), *Biographical Dictionary of Literary Influences: The Nineteenth Century, 1800–1914*, 222–3. Westport, CT: Greenwood.

Mosselmans, B. (2007). *William Stanley Jevons and the Cutting Edge of Economics*. London: Routledge.

Mulberg, J. (1995). *Social Limits to Economic Theory*. London: Routledge.

Murawski, R. (1999). *Recursive Functions and Metamathematics: Problems of Completeness and Decidability, Gödel's Theorems*. Dordrecht: Kluwer.

Musgrave, A. (1981). "'Unreal assumptions' in economic theory: the F-twist untwisted". *Kyklos* 34(3): 377–87.

Nadeau, R. (2003). *The Wealth of Nature: How Mainstream Economics Has Failed the Environment*. New York: Columbia University Press.

Nasar, S. (2011). *Grand Pursuit: The Story of Economic Genius*. New York: Simon & Schuster.

Nelson, R. (2001). *Economics as Religion: From Samuelson to Chicago and Beyond*. University Park, PA: Pennsylvania State University Press.

Neumann, J. von (1947). "The mathematician". In R. Heywood (ed.), *The Works of the Mind*, 180–96. Chicago, IL: University of Chicago Press.

Neumann, J. von & O. Morgenstern (1953). *Theory of Games and Economic Behavior*. Third edn. Princeton, NJ: Princeton University Press.

Neumark, D. (1988). "Employers' discriminatory behavior and the estimation of wage discrimination". *Journal of Human Resources* 23(3): 279–95.

Newton, I. (1848 [1687]). *Newton's Principia: The Mathematical Principles of Natural Philosophy*. Trans. A. Motte. New York: Adee.

Niehans, J. (1990). *A History of Economic Theory: Classic Contributions, 1720–1980*. Baltimore, MA: Johns Hopkins University Press.

Nik-Khah, E. & R. Van Horn (2015). "Inland empire: economics imperialism as an imperative of Chicago neoliberalism". In E. Aydinonat & J. Vromen (eds), *Economics Made Fun: Philosophy of the Pop-Economics*, 71–94. Abingdon: Routledge.

Niiniluoto, I. (2014). "Scientific progress as increasing verisimilitude". *Studies in History and Philosophy of Science, Part A* 46(1): 73–7.

Nobel Committee (1969). "The Sveriges Riksbank Prize in Economic Sciences in Memory of Alfred Nobel 1969 – Ragnar Frisch and Jan Tinbergen". https://www.nobelprize.org/prizes/economic-sciences/1969/summary/.

Nobel Committee (1970). "The Sveriges Riksbank Prize in Economic Sciences in Memory of Alfred Nobel 1970 – Paul A. Samuelson". https://www.nobelprize.org/prizes/economic-sciences/1970/summary/.

Nobel Committee (1982). "The Sveriges Riksbank Prize in Economic Sciences in Memory of Alfred Nobel 1982 – George J. Stigler". https://www.nobelprize.org/prizes/economic-sciences/1982/summary/.

Nordhaus, W. (1991). "To slow or not to slow: the economics of the greenhouse effect". *Economic Journal* 101(407): 920–37.

Nordhaus, W. (1994). *Managing the Global Commons: The Economics of Climate Change*. Cambridge, MA: MIT Press.

North, D. (1961). *The Economic Growth of the United States, 1790–1860*. Hoboken, NJ: Prentice-Hall.

Nowak, L. (1980). *The Structure of Idealization: Towards a Systematic Interpretation of the Marxian Idea of Science*. Dordrecht: Reidel.

Oaxaca, R. & M. Ransom (1994). "On discrimination and the decomposition of wage differentials". *Journal of Econometrics* 61(1): 5–21.

O'Brien, D. (1990). "Lionel Robbins and the Austrian connection". In B. Caldwell (ed.), *Carl Menger and His Legacy in Economics*, 155–84. Durham, NC: Duke University Press.

Olson, M. (1965). *The Logic of Collective Action: Public Goods and the Theory of Groups*. Cambridge, MA: Harvard University Press.

Ostrom, E. (1990). *Governing the Commons: The Evolution of Institutions for Collective Action*. Cambridge: Cambridge University Press.

Otte, M. (2007). "Mathematical history, philosophy and education". *Educational Studies in Mathematics* 66(2): 243–55.

Panteki, M. (2008). "French 'logique' and British 'logic': on the origins of Augustus de Morgan's early logical enquiries, 1805–1835". In D. Gabbay & J. Woods (eds), *Handbook of the History of Logic: Volume 4 – British Logic in the Nineteenth Century*, 381–456. Amsterdam: North-Holland.

Pareto, V. (1896–97). *Cours d'Économie Politique Professé a l'Université de Lausanne, 2 Volumes*. Paris: Pichon, Libraire.

Pareto, V. (1971 [1906]). *Manual of Political Economy*. Trans. A. Schwier. New York: Augustus M. Kelley.

Pareto, V. (1989 [1899]). "Letter to Laurent, 19 January 1899". In V. Pareto, *Letters and Correspondence: The Complete Works of V. Pareto, Volume 30*. Geneva: Droz.

Pareto, V. (1999 [1900]). "On the economic phenomenon: a reply to Benedetto Croce". Trans. F. Priuli, reprinted in J. Wood & M. McClure (eds), *Vilfredo Pareto: Critical Assessments of Leading Economists – Volume I*, 245–61. London: Routledge.

Parisi, F. (2004). "Positive, normative and functional schools in law and economics". *European Journal of Law and Economics* 18(3): 259–72.

Paul, E. (1979). "W. Stanley Jevons: economic revolutionary, political utilitarian". *Journal of the History of Ideas* 40(2): 267–83.

Peart, S. (1996). *The Economics of W. S. Jevons*. London: Routledge.

Peart, S. (2003). "Introduction". In S. Peart (ed.), *W. S. Jevons: Critical Responses – Volume I*, 1–26. London: Routledge.

Peckhaus, V. (2003). "The pragmatism of Hilbert's programme". *Synthese* 137(1/2): 141–56.

Petri, F. (2021). *Microeconomics for the Critical Mind: Mainstream and Heterodox Analysis*. London: Springer.

Phelps, E. (1972). "The statistical theory of racism and sexism". *American Economic Review* 62(4): 659–61.

Phelps Brown, H. (1972). "The underdevelopment of economics". *Economic Journal* 82(325): 1–10.

Pignol, C. (2023). "Economic thought and novels: what can we expect from the economy?". In S. Myrogiannis & C. Repapis (eds), *Economics and Art Theory*, 17–32. Abingdon: Routledge.

Pigou, A. (1912). *Wealth and Welfare*. London: Macmillan.

Pigou, A. (1920). *The Economics of Welfare*. London: Macmillan.

Pizano, D. (2009). *Conversations with Great Economists*. New York: Jórge Pinto Books.

Poisson, S.-D. (1842 [1811]). *A Treatise of Mechanics, 2 Volumes*. Trans. H. Harte. London: Longman.

Pollak, R. & T. Wales (1992). *Demand System Specification and Estimation*. Oxford: Oxford University Press.

Pollini, G. (2001). "The currency and validity of Parsons' interpretation of Vilfredo Pareto's theory of social action". In G. Pollini & G. Sciortino (eds), *Parsons'* The Structure of Social Action *and Contemporary Debates*, 25–44. Milan: Franco Angeli.

Poovey, M. (2008). *Genres of the Credit Economy: Mediating Value in Eighteenth- and Nineteenth-Century Britain*. Chicago, IL: University of Chicago Press.

Popper, K. (1976). "The logic of the social sciences". In T. Adorno (ed.), *The Positivist Dispute in German Sociology*, 87–104. London: Heinemann.

Porter, T. (1986). *The Rise of Statistical Thinking 1820–1900*. Princeton, NJ: Princeton University Press.

Posner, R. (1972). *Economic Analysis of Law*. Boston, MA: Little, Brown.

Powers, C. (2012). "The role of sticking points in Pareto's theory of social systems". In J. Femia & A. Marshall (eds), *Vilfredo Pareto: Beyond Disciplinary Boundaries*, 47–72. Farnham: Ashgate.

Prawitz, D. (1981). "Philosophical aspects of proof theory". In G. Fløistad (ed.), *Contemporary Philosophy: A New Survey*, 235–77. The Hague: Martinus Nijhoff.

Prendergast, C. (2000). *The Triangle of Representation*. New York: Columbia University Press.

Punzo, L. (1991). "The school of mathematical formalism and the Viennese circle of mathematical economists". *Journal of the History of Economic Thought* 13(1): 1–18.

Puttaswamaiah, K. (2002). *Paul Samuelson and the Foundations of Modern Economics*. New Brunswick, NJ: Transaction.

Quetelet, A. (1842). *A Treatise on Man and the Development of his Faculties*. Edinburgh: William & Robert Chambers.

Quiggin, J. (2010). *Zombie Economics: How Dead Ideas Still Walk Among Us*. Princeton, NJ: Princeton University Press.

Quine, W. V. O. (1960). *Word and Object*. Cambridge, MA: MIT Press.

Raatikainen, P. (2003). "Hilbert's program revisited". *Synthese* 137(1/2): 157–77.

Raico, R. (2012). *Classical Liberalism and the Austrian School*. Auburn, AL: Mises Institute.

Raisis, V. (1999). "Expansion and justification of models: the exemplary case of Galileo Galilei". In L. Magnani, N. Nersessian & P. Thagard (eds), *Model-Based Reasoning in Scientific Discovery*, 149–64. New York: Springer.

Ramrattan, L. & M. Szenberg (2019). *American Exceptionalism: Economics, Finance, Political Economy, and Economic Laws*. Basingstoke: Palgrave Macmillan.

Rao, J. (2011). *History of Rotating Machinery Dynamics*. London: Springer.

Rédei, M. (ed.) (2005). *John von Neumann: Selected Letters*. Providence, RI: American Mathematical Society.

Reichenbach, H. (1930). "Tagung für Erkenntnislehre der Exacten Wissenschaften in Königsberg". *Die Naturwissenschaften* 18(50): 1093–4.

Reid, C. (1963). *The New Euclid*. London: Routledge & Kegan Paul.

Reid, C. (1986). *Hilbert-Courant*. New York: Springer.

Reid, C. (1996). *Hilbert*. New York: Springer.

Reill, P. (2005). *Vitalizing Nature in the Enlightenment* Berkeley, CA: University of California Press.

Reis, R. (2018). "Is something really wrong with macroeconomics?". *Oxford Review of Economic Policy* 34(1/2): 132–55.

Reisman, D. (1990). *Alfred Marshall's Mission*. London: Macmillan.

Rescher, N. (2006). *Philosophical Dialectics: An Essay on Metaphilosophy*. Albany, NY: SUNY Press.

Richards, J. (2006). "Historical mathematics in the French eighteenth century". *Isis* 97(4): 700–13.

Riker, W. (1962). *The Theory of Political Coalitions*. New Haven, CT: Yale University Press.

Rima, I. (2012). *Development of Economic Analysis*. Sixth edn. London: Routledge.

Rizvi, A. (2003). "Postwar neoclassical microeconomics". In W. Samuels, J. Biddle & J. Davis (eds), *A Companion to the History of Economic Thought*, 377–94. Oxford: Blackwell.

Robbins, L. (1927). "Mr. Hawtrey on the scope of economics". *Economica* 7(20): 172–8.

Robbins, L. (1932). "Preface to the First Edition". In L. Robbins (1984 [1935]), *An Essay on the Nature and Significance of Economic Science*. Third edn, xli–xliii. London: Macmillan.

Robbins, L. (1933). "Philip Wicksteed as an economist". In L. Robbins (2009 [1970]), *The Evolution of Modern Economic Theory and Other Papers on the History of Economic Thought*, 189–208. New Brunswick, NJ: Transaction.

Robbins, L. (1936). "The place of Jevons in the history of economic thought". In L. Robbins (2009 [1970]), *The Evolution of Modern Economic Theory and Other Papers on the History of Economic Thought*, 169–88. New Brunswick, NJ: Transaction.

Robbins, L. (1938). "Interpersonal comparisons of utility: a comment". *Economic Journal* 48(4): 635–41.

Robbins, L. (1960). "Robertson's *Lectures on Economic Principles*". In L. Robbins (2009 [1970]), *The Evolution of Modern Economic Theory and Other Papers on the History of Economic Thought*, 248–52. New Brunswick, NJ: Transaction.

Robbins, L. (1964). "Bentham in the twentieth century". In L. Robbins (2009 [1970]), *The Evolution of Modern Economic Theory and Other Papers on the History of Economic Thought*, 73–84. New Brunswick, NJ: Transaction.

Robbins, L. (1971). *Autobiography of an Economist*. London: Macmillan.

Robbins, L. (1978). *The Theory of Economic Policy in English Classical Political Economy*. Second edn. London: Macmillan.

Robbins, L. (1979). "Economics and political economy: Richard T. Ely Lecture". Reprinted in L. Robbins (1984 [1935]), *An Essay on the Nature and Significance of Economic Science*. Third edn, xi–xxxiii. London: Macmillan.

Robbins, L. (1984 [1935]). *An Essay on the Nature and Significance of Economic Science*. Third edn. London: Macmillan.

Robbins, L. (2000). *A History of Economic Thought: The LSE Lectures*. Edited by S. Medema & W. Samuels. Paperback edn. Princeton, NJ: Princeton University Press.

Robertson, D. (1976). *A Theory of Party Competition*. London: John Wiley.

Robinson, J. (1962). *Economic Philosophy*. Harmondsworth: Pelican.

Rodgers, W. (ed.) (2006). *Handbook on the Economics of Discrimination*. Cheltenham: Edward Elgar.

Rodin, A. (2014). *Axiomatic Method and Category Theory*. New York: Springer.

Roger, J. (1997). *Buffon: A Life in Natural History*. Trans. S. Bonnefoi. Ithaca, NY: Cornell University Press.

Roncaglia, A. (2005). *The Wealth of Ideas: A History of Economic Thought*. Cambridge: Cambridge University Press.

Rosenberg, A. (1992). *Economics: Mathematical Politics or Science of Diminishing Returns?* Chicago, IL: University of Chicago Press.

Rosenberg, A. (1995). "The metaphysics of microeconomics". *The Monist* 78(3): 352–67.

Ross, D. (2005). *Economic Theory and Cognitive Science: Microexplanation*. Cambridge, MA: MIT Press.

Ross, D. (2012). "The economic agent: not human, but important". In U. Mäki (ed.), *Philosophy of Economics*, 691–735. London: Elsevier.

Rubinstein, A. (2006). "Dilemmas of an economic theorist". *Econometrica* 74(4): 865–83.

Ruccio, D. & J. Amariglio (2003). *Postmodern Moments in Modern Economics*. Princeton, NJ: Princeton University Press.

Rugina, A. (2005). "Nobel Laureate: Paul A. Samuelson (1915–)". Reprinted in J. Wood & M. McClure (eds), *Paul A. Samuelson: Critical Assessments of Contemporary Economists, Volume 3*, 226–58. London: Routledge.

Rutherford, M. (2011). *The Institutionalist Movement in American Economics, 1918–1947: Science and Social Control*. Cambridge: Cambridge University Press.

Ryckman, T. (2016). "What Carnap might have learned from Weyl". In C. Damböck (ed.), *Influences on the Aufbau*, 15–29. New York: Springer.

Saari, D. (1992). "The aggregated excess demand function and other aggregation procedures". *Economic Theory* 2(3): 359–88.

Saatsi, J. (2016). "On the "indispensable explanatory role" of mathematics". *Mind* 125(500): 1045–70.

Saatsi, J. (2017). "Explanation and explanationism in science and metaphysics". In M. Slater & Z. Yudell (eds), *Metaphysics and the Philosophy of Science: New Essays*, 163–93. Oxford: Oxford University Press.

Samuels, W. (2007). "Equilibrium analysis: a middlebrow view". In V. Mosini (ed.), *Equilibrium in Economics: Scope and Limits*, 166–200. London: Routledge.

Samuels, W. (2012). *Pareto on Policy*. Introduction by S. Medema. New Brunswick, NJ: Transaction.

Samuelson, P. (1938). "A note on the pure theory of consumer's behaviour". *Economica* 5(17): 61–71.

Samuelson, P. (1941). "The stability of equilibrium: comparative statics and dynamics". *Econometrica* 9(2): 97–120.

Samuelson, P. (1948). "Consumption theory in terms of revealed preference". *Economica* 15(4): 243–53.

Samuelson, P. (1950). "The problem of integrability in utility theory". *Economica* 17(4): 355–85.

Samuelson, P. (1952). "Economic theory and mathematics – an appraisal". *American Economic Review* 42(2): 56–66.

Samuelson, P. (1955). *Economics: An Introductory Analysis.* Third edn. New York: McGraw Hill.

Samuelson, P. (1961). *Economics: An Introductory Analysis.* Fifth edn. New York: McGraw Hill.

Samuelson, P. (1962). "Economists and the history of ideas". *American Economic Review* 52(1): 1–18.

Samuelson, P. (1963). "Discussion". *American Economic Review* 53(2): 231–6.

Samuelson, P. (1972). "Maximum principles in analytical economics". *American Economic Review* 62(3): 249–62.

Samuelson, P. (1977). "Reaffirming the existence of 'reasonable' Bergson–Samuelson social welfare functions". In K. Crowley (ed.) (1986), *The Collected Scientific Papers of Paul A. Samuelson, Volume V,* 47–54. Cambridge, MA: MIT Press.

Samuelson, P. (1981). "Bergsonian welfare economics". In K. Crowley (ed.) (1986), *The Collected Scientific Papers of Paul A. Samuelson, Volume V,* 3–46. Cambridge, MA: MIT Press.

Samuelson, P. (1983 [1947]). *Foundations of Economic Analysis.* Enlarged edn. Cambridge, MA: Harvard University Press.

Samuelson, P. (1983a). "Introduction to the Enlarged Edition". In P. Samuelson (1983 [1947]), *Foundations of Economic Analysis.* Enlarged edn, xv–xxvi. Cambridge, MA: Harvard University Press.

Samuelson, P. (1983b). "My life philosophy". In K. Crowley (ed.) (1986), *The Collected Scientific Papers of Paul A. Samuelson, Volume V,* 789–96. Cambridge, MA: MIT Press.

Samuelson, P. (1983c). "The 1983 Nobel Prize in Economics". Reprinted in K. Crowley (ed.) (1986), *The Collected Scientific Papers of Paul A. Samuelson, Volume V,* 838–40. Cambridge, MA: MIT Press.

Samuelson, P. (1986). "Economics in my time". In K. Crowley (ed.) (1986), *The Collected Scientific Papers of Paul A. Samuelson, Volume V,* 797–808. Cambridge, MA: MIT Press.

Samuelson, P. (1987). "Out of the closet: a program for the whig history of economic science". *History of Economics Society Bulletin* 9(1): 51–60.

Samuelson, P. (1991). "Conversations with my history-of-economics critics". In G. Shaw (ed.), *Economics, Culture, and Education: Essays in Honour of Mark Blaug,* 3–13. Cheltenham: Edward Elgar.

Samuelson, P. (1992). "The overdue recovery of Adam Smith's reputation as an economic theorist". In M. Fry (ed.), *Adam Smith's Legacy: His Place in the Development of Modern Economics,* 1–15. London: Routledge.

Samuelson, P. (1997). "Credo of a lucky textbook author". *Journal of Economic Perspectives* 11(2): 153–60.

Samuelson, P. (1998a). "How *Foundations* came to be". *Journal of Economic Literature* 36(3): 1375–86.

Samuelson, P. (1998b). "Samuelson, Paul Anthony, as an interpreter of the classical economics". In H. Kurz & S. Neri (eds), *Elgar Companion to Classical Economics, Volume 2*, 329–33. Cheltenham: Edward Elgar.

Samuelson, P. (2004). "Where Ricardo and Mill rebut and confirm arguments of mainstream economists supporting globalization". *Journal of Economic Perspectives* 18(3): 135–46.

Samuelson, P. (2009). "Preface – Thünen: an economist ahead of his time". In J. von Thünen (2009 [1826]), *The Isolated State in Relation to Agriculture and Political Economy, Part III*. Trans. K. Tribe & U. van Suntum, xii–xiv. Basingstoke: Palgrave Macmillan.

Samuelson, P. & S. Swamy (1974). "Invariant economic index numbers and canonical duality: survey and synthesis". *American Economic Review* 64(4): 566–93.

Scarantino, A. (2009). "On the role of values in economic science: Robbins and his critics". *Journal of the History of Economic Thought* 31(4): 449–73.

Scarf, H. (1981). "Comment on: 'On the Stability of Competitive Equilibrium and the Patterns of Initial Holdings: An Example'". *International Economic Review* 22(2): 469–70.

Schabas, M. (1990). *A World Ruled by Number: William Stanley Jevons and the Rise of Mathematical Economics*. Princeton, NJ: Princeton University Press.

Schabas, M. (2009). "Constructing 'the economy'". *Philosophy of the Social Sciences* 39(1): 3–19.

Scheibe, E. (2022). *The Reduction of Physical Theories: A Contribution to the Unity of Physics – Part 1: Foundations and Elementary Theory*. Trans. B. Falkenburg & G. Jaeger. New York: Springer.

Schmidt, G. (2011). *Rational Mathematics*. Cambridge: Cambridge University Press.

Schröder, E. (1873). *Lehrbuch der Arithmetik und Algebra*. Leipzig: Teubner.

Schumpeter, J. (1984 [1954]). *History of Economic Analysis*. Oxford: Oxford University Press.

Schwarze, R. (2001). *Law and Economics of International Climate Change Policy*. New York: Springer.

Scott, A. (2007). *The Nonlinear Universe: Chaos, Emergence, Life*. New York: Springer.

Screpanti, E. & S. Zamagni (2005). *An Outline of the History of Economic Thought*. Second edn. Trans. D. Field & L. Kirby. Oxford: Oxford University Press.

Sekerler Richiardi, P. (2011). "Is Jevons a liberal of happiness?". In R. Ece & H. Igersheim (eds), *Freedom and Happiness in Economic Thought and Philosophy: From Clash to Reconciliation*, 85–101. London: Routledge.

Sen, A. (1973). "Behaviour and the concept of preference". *Economica* 40(3): 241–59.

Sen, A. (1980). "Description as choice". *Oxford Economic Papers* 32(3): 353–69.

Sent, E.-M. (1998). "Engineering dynamic economics". In J. Davis (ed.), *New Economics and Its History*, 41–61. Durham, NC: Duke University Press.

Sent, E.-M. (1999). "Methodology in economics". In P. O'Hara (ed.), *Encyclopedia of Political Economy: Volume 2, L–Z*, 727–31. London: Routledge.

Sentilles, D. (1975). *A Bridge to Advanced Mathematics*. Baltimore, MA: Williams & Wilkins.

Shackle, G. (1967). *The Years of High Theory: Invention and Tradition in Economic Thought, 1926–1939*. Cambridge: Cambridge University Press.

Shafer, W. & H. Sonnenschein (1982). "Market excess demand functions". In K. Arrow & M. Intriligator (eds), *Handbook of Mathematical Economics: Volume II*, 671–93. Amsterdam: North-Holland.

Shankar, N. (1994). *Metamathematics, Machines and Gödel's Proof*. Cambridge: Cambridge University Press.

Shapiro, S. (2009). "Categories, structures, and the Frege–Hilbert controversy: the status of meta-mathematics". In S. Lindström *et al.* (eds), *Logicism, Intuitionism, and Formalism: What Has Become of Them?*, 435–48. Berlin: Springer.

Shavell, S. (2004). *Foundations of Economic Analysis and Law*. Cambridge, MA: Harvard University Press.

Shiller, R. (2012). *Finance and the Good Society*. Princeton, NJ: Princeton University Press.

Shubik, M. (1977). "Competitive and controlled price economies: the Arrow–Debreu model revisited". In G. Schwodiauer (ed.), *Equilibrium and Disequilibrium in Economic Theory*, 213–24. Dordrecht: Reidel.

Sieg, W. (1988). "Hilbert's program sixty years later". *Journal of Symbolic Logic* 53(2): 338–48.

Sieg, W. (2012). "In the shadow of incompleteness: Hilbert and Gentzen". In P. Dybjer *et al.* (eds), *Epistemology versus Ontology: Essays on the Philosophy and Foundations of Mathematics in Honour of Per Martin-Löf*, 87–128. New York: Springer.

Sieg, W. (2013). *Hilbert's Programs and Beyond*. Oxford: Oxford University Press.

Sieg, W. & M. Ravaglia (2005). "David Hilbert and Paul Bernays, *Grundlagen der Mathematik*, First Edition (1934, 1939)". In I. Grattan-Guinness (ed.), *Landmark Writings in Western Mathematics 1640–1940*, 981–98. London: Elsevier.

Siegel, D. (1991). *Innovation in Maxwell's Electromagnetic Theory: Molecular Vortices, Displacement Current, and Light*. Cambridge: Cambridge University Press.

Sigot, N. (2002). "Jevons's debt to Bentham: mathematical economy, morals and psychology". *The Manchester School* 70(2): 262–78.

Silberberg, E. (2005). "Foreword". In M. Caputo (2005), *Foundations of Dynamic Economic Analysis: Optimal Control Theory and Applications*, vii. Cambridge: Cambridge University Press.

Simpson, S. (1988). "Partial realizations of Hilbert's program". *Journal of Symbolic Logic* 53(2): 349–63.

Skousen, M. (1997). "The perseverance of Paul Samuelson's *Economics*". *Journal of Economic Perspectives* 11(2): 137–52.

Slater, M. (1950). "Lagrange multipliers revisited". Cowles Commission Discussion Paper, Mathematics 403.

Smith, P. (2013). *An Introduction to Gödel's Theorems*. Second edn. Cambridge: Cambridge University Press.

Smith, R. (1977). "The human significance of biology: Carpenter, Darwin, and the *Vera Causa*". In U. Knoepflmacher & G. Tennyson (eds), *Nature and the Victorian Imagination*, 216–30. Berkeley, CA: University of California Press.

Smith, R. (1981). *Trial by Medicine: Insanity and Responsibility in Victorian Trials*. Edinburgh: Edinburgh University Press.

Smoryński, C. (1977). "The incompleteness theorems". In J. Barwise (ed.), *Handbook of Mathematical Logic*, 821–66. Amsterdam: North-Holland.

Smullyan, R. (1992). *Gödel's Incompleteness Theorems*. Oxford: Oxford University Press.

Sonnenschein, H. (1972). "Market excess demand functions". *Econometrica* 40(3): 549–63.

Sonnenschein, H. (1973). "Do Walras' identity and continuity characterize the class of community excess demand functions?". *Journal of Economic Theory* 6(4): 345–54.

Soto, J. de (2009). *The Theory of Dynamic Efficiency*. London: Routledge.

Souter, R. (1933a). "'The Nature and Significance of Economic Science' in recent discussion". *Quarterly Journal of Economics* 47(3): 377–413.

Souter, R. (1933b). *Prolegomena to Relativity Economics: An Elementary Study in the Mechanics and Organics of an Expanding Economic Universe*. New York: Columbia University Press.

Spengler, J. (1934). "Have values a place in economics?". *International Journal of Ethics* 44(3): 313–31.

Spiegler, P. (2005). On a Difference Between Robbins and Jevons Regarding the Proper Scope of Economics. Unpublished manuscript, Harvard University.

Stedall, J. (2011). *From Cardano's Great Art to Lagrange's Reflections: Filling a Gap in the History of Algebra*. Zurich: European Mathematical Society.

Stedall, J. (2012). *The History of Mathematics: A Very Short Introduction*. Oxford: Oxford University Press.

Steedman, I. (1989). "Rationality, economic man and altruism in P. H. Wicksteed's *Common Sense of Political Economy*". Reprinted in S. Zamagni (ed.) (1995), *The Economics of Altruism*, 108–22. Cheltenham: Edward Elgar.

Stewart, I. & D. Tall (2015). *The Foundations of Mathematics*. Second edn. Oxford: Oxford University Press.

Stigler, G. (1971). "The theory of economic regulation". *Bell Journal of Economics and Management Science* 2(1): 3–21.

Stigler, G. (1972). "Economic competition and political competition". *Public Choice* 13(Fall): 91–106.

Stigler, G. (1976). "The Xsistence of X-efficiency". *American Economic Review* 60(1): 213–16.

Stigler, G. (1984). "Economics: the imperial science?". *Scandinavian Journal of Economics* 86(3): 301–13.

Stigler, G. & G. Becker (1977). "De gustibus non est disputandum". *American Economic Review* 67(2): 76–90.

Stigler, S. (1982). "Jevons as statistician". *The Manchester School* 50(4): 354–65.

Stigler, S. (1999). *Statistics on the Table: The History of Statistical Concepts and Methods*. Cambridge, MA: Harvard University Press.

Stoeger, H. (2011). "The troubled genius: myths and facts". In A. Ziegler & C. Perleth (eds), *Excellence: Essays in Honour of Kurt A. Heller*, 76–93. Zurich: Lit Verlag.

Strassman, D. & L. Polanyi (1995). "The economist as storyteller: what the texts reveal". In S. Feiner *et al.* (eds), *Out of the Margin: Feminist Perspectives on Economic Theory*, 94–106. London: Routledge.

Suárez, M. (2010). "Scientific representation". *Philosophy Compass* 5(1): 91–101.

Sugden, R. (2000). "Credible worlds: the status of theoretical models in economics". *Journal of Economic Methodology* 7(1): 1–31.

Sugden, R. (2009a). "Credible worlds, capacities and mechanisms". *Erkenntnis* 70(1): 3–27.

Sugden, R. (2009b). "Can economics be founded on 'indisputable facts of experience'? Lionel Robbins and the pioneers of neoclassical economics". *Economica* 76 (Supp.): 857–72.

Sugden, R. (2011). "Explanations in search of observations". *Biology and Philosophy* 26(5): 717–36.

Swedberg, R. (1990). *Economics and Sociology: Redefining their Boundaries. Conversations with Economists and Sociologists.* Princeton, NJ: Princeton University Press.

Szász, D. (2011). "John von Neumann, the mathematician". *The Mathematical Intelligencer* 33(2): 42–51.

Szreter, S. (1996). *Fertility, Class and Gender in Britain 1860–1940.* Cambridge: Cambridge University Press.

Tappenden, J. (2013). "The mathematical and logical background to analytic philosophy". In M. Beaney (ed.), *Oxford Handbook of the History of Analytic Philosophy*, 318–54. Oxford: Oxford University Press.

Tarascio, V. (1972). "Vilfredo Pareto and marginalism". *History of Political Economy* 4(2): 406–25.

Tarascio, V. (1974). "Pareto on political economy". *History of Political Economy* 6(4): 361–80.

Tasić, V. (2001). *Mathematics and the Roots of Postmodern Thought.* Oxford: Oxford University Press.

Thiele, R. (2005). "Hilbert and his twenty-four problems". In G. Van Brummelen & M. Kinyon (eds), *Mathematics and the Historian's Craft: The Kenneth O. May Lectures*, 243–95. New York: Springer.

Thompson, G. (1999). "Strategy and tactics in the pedagogy of economics: what should be done about neoclassical economics?". In R. Garnett (ed.), *What Do Economists Know? New Economics of Knowledge*, 223–35. London: Routledge.

Thweatt, W. (1983). "Origins of the terminology 'supply and demand'". *Scottish Journal of Political Economy* 30(3): 287–94.

Tinbergen, J. (1935). "Quantitative Fragen der Konjunkturpolitik". *Weltwirtschaftliches Archiv* 42(1): 366–99.

Tinbergen, J. (1937). *An Econometric Approach to Business Cycle Problems.* Paris: Hermann & Cie.

Toepell, M. (2005). "David Hilbert, *Grundlagen der Geometrie*, First Edition (1899)". In I. Grattan-Guinness (ed.), *Landmark Writings in Western Mathematics 1640–1940*, 710–23. London: Elsevier.

Tol, R. (2009). "The economic effects of climate change". *Journal of Economic Perspectives* 23(2): 29–51.

Tomalin, M. (2006). *Linguistics and the Formal Sciences: The Origins of Generative Grammar.* Cambridge: Cambridge University Press.

Toynbee, A. (2011 [1884]). *Lectures on the Industrial Revolution in England: Popular Addresses, Notes and Other Fragments.* Cambridge: Cambridge University Press.

Tribe, K. (2007). "Historical schools of economics: German and English". In W. Samuels, J. Biddle & J. Davis (eds), *A Companion to the History of Economic Thought*, 215–30. Paperback edn. Oxford: Blackwell.

Tribe, K. (2015). *The Economy of the Word: Language, History, and Economics.* Oxford: Oxford University Press.

Trupiano, J. (2013). "Jevons, William Stanley". In D. Dieterle (ed.), *Economic Thinkers: A Biographical Encyclopedia*, 159–61. Santa Barbara, CA: Greenwood.

Udehn, L. (1992). "The limits of economic imperialism". In U. Himmelstrand (ed.), *Interfaces in Economic and Social Analysis*, 239–80. London: Routledge.

Udehn, L. (2001). *Methodological Individualism: Background, History and Meaning*. London: Routledge.

Uzawa, H. (1960). "Preferences and rational choice in the theory of consumption". In K. Arrow, S. Karlin & P. Suppes (eds), *Proceedings of the First Stanford Symposium on Mathematical Methods in the Social Sciences*, 129–48. Stanford, CA: Stanford University Press.

Vaisman, I. (1980). *Foundations of Three-Dimensional Euclidean Geometry*. New York: Marcel Dekker.

Van Bouwel, J. (2011). "An atlas for the social world: what should it (not) look like? Interdisciplinarity and pluralism in the social sciences". In D. Aerts *et al.* (eds), *Worldviews, Science and Us: Interdisciplinary Perspectives on Worlds, Cultures and Society*, 43–72. Singapore: World Scientific.

Varian, H. (1984). "Gerard Debreu's contributions to economics". *Scandinavian Journal of Economics* 86(1): 4–14.

Vaughan, R. (2016). "Goldbach's conjectures: a historical perspective". In J. Nash & M. Rassias (eds), *Open Problems in Mathematics*, 479–520. Boston, MA: Springer.

Vaughn, K. (1994). *Austrian Economics in America: The Migration of a Tradition*. Cambridge: Cambridge University Press.

Wald, A. (1951 [1936]). "On some systems of equations of mathematical economics". Trans. from the original 1936 German version by O. Eckstein. *Econometrica* 19(4): 368–403.

Walker, D. (1987). "Walras's theories of tatonnement". *Journal of Political Economy* 95(4): 758–74.

Walker, D. (2005). *Walras's Market Models*. Cambridge: Cambridge University Press.

Walker, D. (2011). *Walrasian Economics*. Cambridge: Cambridge University Press.

Walras, L. (1984/1954 [1874]). *Elements of Pure Economics: Or the Theory of Social Wealth*. Trans. W. Jaffé. Philadelphia, PA: Orion Editions.

Wang, H. (1981). "Some facts about Kurt Gödel". *Journal of Symbolic Logic* 46(3): 653–9.

Warsh, D. (1993). *Economic Principals: Masters and Mavericks of Modern Economics*. New York: Free Press.

Watson, M. (2014). *Uneconomic Economics and the Crisis of the Model World*. Basingstoke: Palgrave.

Watson, M. (2018). *The Market*. Newcastle upon Tyne: Agenda Publishing.

Webb, J. (1997). "Hilbert's formalism and arithmetization of mathematics". *Synthese* 110(1): 1–14.

Weber, C. (2001). "Pareto and the 53 percent ordinal theory of utility". *History of Political Economy* 33(3): 541–76.

Weil, A. (1978). "Who betrayed Euclid? (Extract from a Letter to the Editor)". *Archive for History of Exact Sciences* 19(2): 91–3.

Weil, A. (1992). *The Apprenticeship of a Mathematician*. Boston, MA: Birkhäuser.

Weintraub, R. (1977). "The microfoundations of macroeconomics: a critical survey". *Journal of Economic Literature* 15(1): 1–23.

Weintraub, R. (1985). *General Equilibrium Analysis: Studies in Appraisal*. Cambridge: Cambridge University Press.

Weintraub, R. (1998). "Axiomatisches Mißverständnis". *Economic Journal* 108(451): 1837–47.

Weintraub, R. (2002). *How Economics Became a Mathematical Science*. Durham, NC: Duke University Press.

Weintraub, R. & T. Gayer (2000). "Negotiating at the boundary: Patinkin vs Phipps". *History of Political Economy* 32(3): 441–71.

Weintraub, R. & T. Gayer (2001). "Equilibrium proofmaking". *Journal of the History of Economic Thought* 23(4): 421–42.

Weintraub, R. & P. Mirowski (1994). "The pure and the applied: Bourbakism comes to mathematical economics". *Science in Context* 7(2): 245–72.

Weisberg, M. (2006). "Robustness analysis". *Philosophy of Science* 73(5): 730–42.

Weisberg, M. (2007). "Who is a modeler?". *British Journal for the Philosophy of Science* 58(2): 207–33.

Weisberg, M. (2013). *Simulation and Similarity: Using Models to Understand the World*. Oxford: Oxford University Press.

Weyl, H. (1944). "David Hilbert and his mathematical work". *Bulletin of the American Mathematical Society* 50(9): 612–54.

Whaples, R. (1991). "A quantitative history of the *Journal of Economic History* and the cliometric revolution". *Journal of Economic History* 51(2): 289–301.

Whewell, W. (1858). *Novum Organon Renovatum. Being the Second Part of the Philosophy of the Inductive Sciences*. London: John W. Parker.

White, M. (1994). "The moment of Richard Jennings: the production of Jevons's marginalist economic agent". In P. Mirowski (ed.), *Natural Images in Economic Thought: "Markets Read in Tooth and Claw"*, 197–230. Cambridge: Cambridge University Press.

White, M. (1996). "No matter of regret: the Cambridge critique(s) of Jevons's 'hedonics'". In P. Groenewegen (ed.), *Economics and Ethics?*, 103–20. London: Routledge.

White, M. (2002). "Manufacturing the Smithian paradox of value". In C. Sardini & P. Kriesler (eds), *Keynes, Post-Keynesianism and Political Economy: Essays in Honour of Geoff Harcourt, Volume Three*, 3–22. London: Routledge.

White, M. (2004). "A grin without a cat: W. S. Jevons' elusive equilibrium". In T. Aspromourgos & J. Lodewijks (eds), *History and Political Economy: Essays in Honour of P. D. Groenewegen*, 88–106. London: Routledge.

Wicksteed, P. (2003 [1910]). *The Common Sense of Political Economy and Selected Papers and Reviews on Economic Theory*, two volumes. London: Routledge.

Wicksteed, P. (2014 [1888]). *The Alphabet of Economic Science: Elements of the Theory of Value or Worth*. Charleston, SC: Nabu Press.

Wilson, D. & W. Dixon (2012). *A History of Homo Economicus: The Nature of the Moral in Economic Theory*. London: Routledge.

Wimsatt, W. (1987). "False models as a means to truer theories". In M. Nitecki & A. Hoffmann (eds), *Neutral Models in Biology*, 23–55. Oxford: Oxford University Press.

Winch, D. (1973). "Marginalism and the boundaries of economic science". In R. Black, A. Coats & C. Goodwin (eds), *The Marginal Revolution in Economics: Interpretation and Evaluation*, 59–77. Durham, NC: Duke University Press.

Winch, D. (2002). "Does progress matter?". In S. Boehm *et al.* (eds), *Is There Progress in Economics? Knowledge, Truth and the History of Economic Thought*, 3–20. Cheltenham: Edward Elgar.

Winther, R. (2006). "On the dangers of making scientific models ontologically independent: taking Richard Levins' warnings seriously". *Biology and Philosophy* 21(5): 703–24.

Witt, U. (2009). "Novelty and the bounds of unknowledge in economics". *Journal of Economic Methodology* 16(4): 361–75.

Wittman, D. (1973). "Parties as utility maximizers". *American Political Science Review* 67(2): 490–98.

Wong, S. (2006 [1978]). *Foundations of Paul Samuelson's Revealed Preference Theory: A Study by the Method of Rational Reconstruction*. Revised edn. London: Routledge.

Wood, J. & M. McClure (1999). "Introduction". In J. Wood & M. McClure (eds), *Vilfredo Pareto: Critical Assessments of Leading Economists, Volume I*. London: Routledge.

Woodward, J. (2006). "Some varieties of robustness". *Journal of Economic Methodology* 13(2): 219–40.

Wootton, B. (1938). *Lament for Economics*. London: Allen & Unwin.

Yandell, B. (2010). "David Hilbert". In T. Gowers, J. Barrow-Green & I. Leader (eds), *Princeton Companion to Mathematics*, 788–9. Princeton, NJ: Princeton University Press.

Zach, R. (2007). "Hilbert's program then and now". In D. Jacquette (ed.), *Philosophy of Logic*, 411–48. Boston, MA: Elsevier.

Zamarovský, P. (2017). "Epistemology and the transformation of knowledge in the global age: God and the epistemology of mathematics". In Z. Delić (ed.), *Epistemology and Transformation of Knowledge in the Global Age*, 85–102. Rijeka: InTech.

Zerbe, R. (2001). *Economic Efficiency in Law and Economics*. Cheltenham: Edward Elgar.

Zouboulakis, M. (2014). *The Varieties of Economic Rationality: From Adam Smith to Contemporary Behavioural and Evolutionary Economics*. Abingdon: Routledge.

Index